213
Advances in Polymer Science

Advances in Polymer Science
Recently Published and Forthcoming Volumes

Photoresponsive Polymers I

Volume Editors: Seth R. Marder · Kwang-Sup Lee

With contributions by
S. Barlow · K. D. Belfield · M. V. Bondar · S. Juodkazis
S. R. Marder · H. Misawa · V. Mizeikis · J. W. Perry
M. Rumi · J. Wang · S. Yao

 Springer

The series *Advances in Polymer Science* presents critical reviews of the present and future trends in polymer and biopolymer science including chemistry, physical chemistry, physics and material science. It is adressed to all scientists at universities and in industry who wish to keep abreast of advances in the topics covered.

As a rule, contributions are specially commissioned. The editors and publishers will, however, always be pleased to receive suggestions and supplementary information. Papers are accepted for *Advances in Polymer Science* in English.

In references *Advances in Polymer Science* is abbreviated *Adv Polym Sci* and is cited as a journal.

Springer WWW home page: springer.com
Visit the APS content at springerlink.com

ISBN 978-3-540-69448-9   e-ISBN 978-3-540-69450-2
DOI 10.1007/978-3-540-69450-2

Advances in Polymer Science ISSN 0065-3195

Library of Congress Control Number: 2008932949

Cover design: WMXDesign GmbH, Heidelberg
Typesetting and Production: le-tex publishing services oHG, Leipzig

Printed on acid-free paper

9 8 7 6 5 4 3 2 1 0

springer.com

Advances in Polymer Science
Also Available Electronically

For all customers who have a standing order to Advances in Polymer Science, we offer the electronic version via SpringerLink free of charge. Please contact your librarian who can receive a password or free access to the full articles by registering at:

springerlink.com

If you do not have a subscription, you can still view the tables of contents of the volumes and the abstract of each article by going to the SpringerLink Homepage, clicking on "Browse by Online Libraries", then "Chemical Sciences", and finally choose Advances in Polymer Science.

You will find information about the

– Editorial Board
– Aims and Scope
– Instructions for Authors
– Sample Contribution

at springer.com using the search function.

Color figures are published in full color within the electronic version on SpringerLink.

Preface

Over the past 25 years or so there has been a revolution in the development of functional polymers. While many polymers as commodities represent huge markets, new materials with a high degree of functionality have been developed. Such specialty polymers play important roles in our day-to-day lives. The current volumes 213 and 214 of Advances in Polymer Science focus on photoresponsive polymers. In particular polymers that can either change the properties of a beam of light that passes through them or who change their properties in response to light. Volume 213 starts with an introduction to two-photon absorption by Rumi, Barlow, Wang, Perry, and Marder. In this chapter they develop the basic concepts of two-photon absorption, and describe structure–property relationships for a variety of symmetrical and unsymmetrical molecules. The applications of these materials in 3D microfabrication of polymers, metals, and oxide materials are detailed in the chapter entitled "Two-Photon Absorber and Two-Photon Induced Chemistry" contributed by the same group of authors. Then Belfield, Bondar, and Yao describe the molecules, dendrimers, oligomers, and polymers that can be excited by two-photon absorption and their application in processing materials with three-dimensional spatial control in their chapter entitled "Two-Photon Absorbing Photonic Materials." Specifically they describe the development of symmetrical and polar conjugated materials for two-photon absorption and their use as photo-initiators for 3D microfabrication. Juodkazis, Mizeikis, and Misawa also explore multiphoton processing of materials in their chapter, and provide more focus on the processing aspects of these materials and discuss the state-of-the-art in resolution.

In Volume 214, Hoppe and Sariciftci describe how organic semiconducting polymers can be used to produce electrical power when excited by light in the chapter entitled "Polymeric Photovoltaic Devices." In particular the authors review approaches based upon blends of conjugated polymers with small molecules that are approaching a point where they can be considered for commercialization. This is followed by a chapter by McGrath and D'Ambruoso entitled "Energy Harvesting in Synthetic Dendritic Materials" where they describe dendritic materials that can absorb light across various parts of the UV–visible spectrum and funnel energy down to a low energy absorber, which can be useful for a variety of applications including photovoltaics. Finally,

Baldeck and Andraud provide a chapter entitled "Exitonic Coupled Oligomers and Dendrimers for Two-Photon Absorption," wherein the concepts of excitonic coupling are developed and their relevance to multi-photon absorption processes are described.

The editors hope that these volumes will provide the reader with an overview of various aspects of photoresponsive polymers. We recommend that readers also examine other volumes in this series to learn more about related topics. In addition the editors thank the authors of the chapters in these volumes and the staff of Springer for their contribution to these volumes and accept responsibility for any errors or inaccuracies.

Atlanta & Daejeon, May 2008   S. R. Marder and K.-S. Lee

Contents

Contents of Volume 214

Photoresponsive Polymers II

Volume Editors: Marder, S. R., Lee, K.-S.
ISBN: 978-3-540-69452-6

Adv Polym Sci (2008) 213: 1–95
DOI 10.1007/12_2008_133
© Springer-Verlag Berlin Heidelberg
Published online: 23 May 2008

Two-Photon Absorbing Materials and Two-Photon-Induced Chemistry

Mariacristina Rumi (✉) · Stephen Barlow · Jing Wang · Joseph W. Perry · Seth R. Marder

School of Chemistry and Biochemistry
and Center for Organic Photonics and Electronics,
Georgia Institute of Technology, Atlanta, Georgia 30332-0400, USA
mrumi@gatech.edu

Abstract Two-photon absorption, the process by which an excited molecule or material is produced by the simultaneous absorption of two photons, has been studied extensively in recent years, from both fundamental and application points of view. On one side, the field has been expanded with the measurement of two-photon absorption spectra and cross sections for a wide range of conjugated molecules. In this contribution, we will review the two-photon properties of some of these classes of materials, and we will discuss structure/property relationships that have been developed from these investigations. On the other side, two-photon absorption has been exploited as a means to activate a variety of chemical and physical processes with sub-diffraction–limited resolution, because three-dimensional spatial confinement of the excitation volume in a material irradiated by a tightly focused laser beam can be achieved via two-photon or, more generally, multiphoton absorption processes. This characteristic has led to applications in a number of technological fields, such as microfabrication and laser scanning fluorescence microscopy. Here, we will survey material systems that have been developed to activate radical or cationic polymerization reactions, deprotection of functional groups, and singlet oxygen generation via two-photon excitation of one of the components in the system.

Keywords Structure/property relationships · Two-photon absorption ·
Two-photon chromophores · Two-photon deprotection ·
Two-photon-induced chemistry · Two-photon polymerization

1
Introduction

Over the last 15 years, interest in two-photon absorption (2PA) has been revitalized by the demonstration that some aspects of the two-photon absorption process, specifically the fact that the excitation produced by a tightly focused laser beam in a 2PA active material can be confined in three dimensions (3D), could be exploited in practical applications with far-reaching technological potential. This volume confinement translates, for example, into the ability to probe or modify a specific point inside a material with 3D control over its position with sub-diffraction limited resolution. For example, Rentzepis et al. [1] and Strickler et al. [2] applied the localization principle to the creation of 3D optical memories, in which multiple layers in a material can be independently used to store and retrieve information, with the potential to achieve overall data density on the order of 10^{12} bits/cm^3. Similarly, chemical modifications occurring selectively in the excited volume can be used to sequentially write patterns and features in the material, while not affecting the unexposed sections. Webb and collaborators have shown that "objects" can be fabricated in this way with lateral dimension on the micron scale [3, 4]. Webb and collaborators have also exploited the localization of the excitation as an imaging tool, by collecting the fluorescence emitted by the selected volume element [3, 5]. Optical microscopes based on the imaging of the fluorescence emitted after 2PA are now commercially available and the technique has since become a powerful new tool in the field of cell and molecular biology.

In this chapter, we will first provide an introduction to the process of 2PA (Sect. 2) and then a description of the 2PA spectroscopy of various classes of small organic molecules (Sect. 3). We will then survey the basic structure/property relationships for 2PA in π-conjugated molecules and how these can be exploited for the design of new materials with enhanced 2PA cross sections (Sect. 4). Finally, we will provide a brief overview of chemical processes which can be enabled by 2PA (Sect. 5). This chapter is not intended to be an exhaustive review of the two-photon absorbers which have been investigated or of the two-photon microfabrication literature. Instead, this chapter is an introduction to the materials and chemistries used in applications based on two-photon absorption, providing context to several other chapters in this volume. In particular, the work of Belfield, Baldeck, and Andraud is discussed in detail in other chapters of this volume. Other reviews on 2PA chromophores [6], microfabrication [7–9], and two-photon fluorescence microscopy [10, 11] can be found in the recent literature.

2
The Two-Photon Absorption Process

Two-photon absorption is a process in which a molecule or material simultaneously absorbs two photons [12]. In the case of degenerate (or one-color) 2PA, the two photons have the same energy, which is approximately half the energy (twice the wavelength) normally required to reach an excited state in the same material. This is illustrated schematically in Fig. 1, where g is the ground state of a molecule, e is the lowest excited state, and the dashed lines represent the manifold of rovibrational levels in each electronic state. In general, a molecule can be excited to state e by absorbing a photon (one-photon absorption, 1PA) with energy equal to the energy difference between states g and e (thick arrow in the figure), assuming that this transition is allowed by the selection rules of the process. In the case illustrated in the left panel of the figure, the same state can be reached through the absorption of

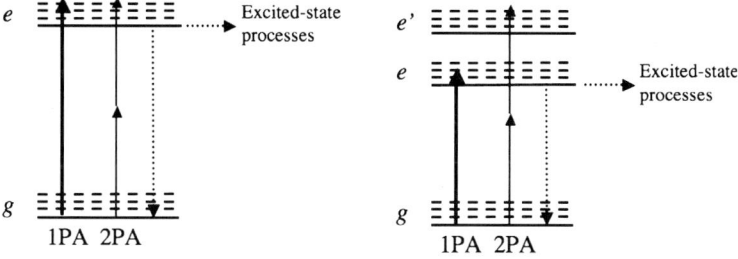

Fig. 1 Energy diagrams illustrating the process of one-photon absorption (1PA) and two-photon absorption (2PA) for: (*left*) a molecule without an inversion center; (*right*) a molecule with an inversion center. The length of each *arrow* is proportional to the photon energy. The *dotted arrows* represent possible deexcitation pathways

two identical photons, each with half the energy of the spacing between the states. This description applies to the behavior of many molecules without an inversion center. However, because the selection rules for 1PA and 2PA are different [13, 14], if a molecule is centrosymmetric, a 1PA-allowed state e cannot also be reached directly by 2PA. In these systems, the lowest electronic state accessible through 2PA usually lies at higher energy than the 1PA-allowed state e (state e' in the right panel of Fig. 1). After being excited by either 1PA or 2PA, a molecule typically undergoes rapid (ps timescale) internal conversion until the system reaches the lowest level of the e manifold. From this level, the system can relax back to the ground state radiatively or nonradiatively (dotted downward arrows in the figure), typically on a ns timescale [15]. Additionally, the molecule can lose the excess energy though other processes that can take place from the excited state, for example, energy transfer or electron transfer, or can react with other molecular species in its surroundings. In Sect. 5, we will discuss some examples of chemical processes that have been accomplished using 2PA.[1] In general, in centrosymmetric molecules, there is an exclusivity between the states that can be reached directly by 1PA and 2PA from the ground state. In fact, only transitions between states with different parity with respect to the inversion center (that is transition between a state of *gerade* and a state of *ungerade* parity, or vice versa), can be one-photon allowed in centrosymmetric molecules, while 2PA can only occur between states with the same parity (from *gerade* to *gerade*, or from *ungerade* to *ungerade*) [14, 16]. For molecules with high symmetry, some transitions are neither one-photon nor two-photon allowed. For noncentrosymmetric molecules, the selection rules for 1PA and 2PA are not mutually exclusive, and excited states exist that can be reached by either type of transition from the ground state (although, depending on the symmetry, some states may still not be allowed for either or both of the processes). Overall, within the dipole approximation and for molecules with a totally symmetric ground state, 2PA transitions from the ground state can access only excited states with the same symmetry as one of the products $i \times j$, where $i, j = x, y, z$ (this selection rule is reminiscent of that for Raman activity).

The probability of a molecule absorbing one photon is proportional to the intensity of the excitation beam:

$$n^{(1)} = \sigma(\nu)N_g \frac{I}{h\nu} , \tag{1}$$

where $n^{(1)}$ is the number of molecules excited by 1PA per unit time and unit volume in the material, σ is the cross section of the absorption process at frequency ν, N_g is the density of molecules in the ground state g, I is the in-

---

[1] For most molecules, the energy level reached after internal conversion is the same irrespective of the order of the absorption process used to generate the excited species. As a consequence, all photo-induced processes traditionally activated by 1PA can also be accomplished using 2PA.

tensity of the excitation source (in energy per unit time and area), and $h\upsilon$ is the photon energy.

In contrast, the probability of a molecule absorbing two photons simultaneously is proportional to the square of the intensity of the excitation beam:

$$n^{(2)} = \frac{1}{2}\delta(\upsilon)N_g\left(\frac{I}{h\upsilon}\right)^2 , \qquad (2)$$

where $n^{(2)}$ is now the number of molecules excited by 2PA in the unit volume per unit time and $\delta(\upsilon)$ is the 2PA cross section for a photon of energy $h\upsilon$. The prefactor of 1/2 reflects the fact that two photons are needed to excite one molecule.[2]

As a result of this intensity dependence, 2PA provides a mechanism by which chemical or physical processes can be activated with high spatial resolution in three dimensions, with excitation being confined to widths down to $\sim$60 nm. The ability of 2PA to excite molecules with this 3D spatial resolution results from the fact that the intensity of a focused laser beam decreases approximately as the square of the distance, z, from the focus. Thus, since 2PA scales quadratically with light intensity (see Eq. 2), the number of excited states formed by 2PA is proportional to z^{-4}, whereas the number of excited states formed by one-photon absorption is proportional to z^{-2}. This is illustrated in Fig. 2, where the dependence of the intensity and of the square of the intensity for a laser beam with a Gaussian profile is graphed as a function of z, the distance from the focal plane. As mentioned above, the 2PA transition requires photons with approximately half the energy (twice the wavelength) of that of the photons needed for 1PA into the lowest excited state of a molecule. If the beams at wavelengths λ and 2λ have the same waist at the focus, ω_0, the excitation rate follows the trend of the solid line in the figure for the 1PA case and the dashed line in the 2PA case. It can be seen that the width of the peak is narrower in the latter situation.[3] The much stronger distance dependence of 2PA means that excitation can be essentially confined to a volume on the order of λ^3, where λ is the wavelength of the excitation light.

The 2PA process is very weak relative to one-photon excitation, in the sense that the ratio $n^{(2)}/n^{(1)}$ (from Eqs. 1 and 2) is typically small for intensities below about 10 GW/cm^2 and for the absorption cross sections seen in typical

---

[2] However, the use of the 1/2 factor is not universal and some authors define the 2PA cross section based on the equation:

$$n^{(2)} = \delta^* N_g\left(\frac{I}{h\upsilon}\right)^2 .$$

[3] In practical applications, it should be kept in mind that the beam waist itself usually depends on the wavelength (at the diffraction limit, ω_0 is proportional to λ) and, therefore, that the comparison between 1PA and 2PA excitation rates may not be as straightforward as that shown in Fig. 2. In general, the 2PA excitation volume depends on the focusing conditions and beam parameters used for the material excitation.

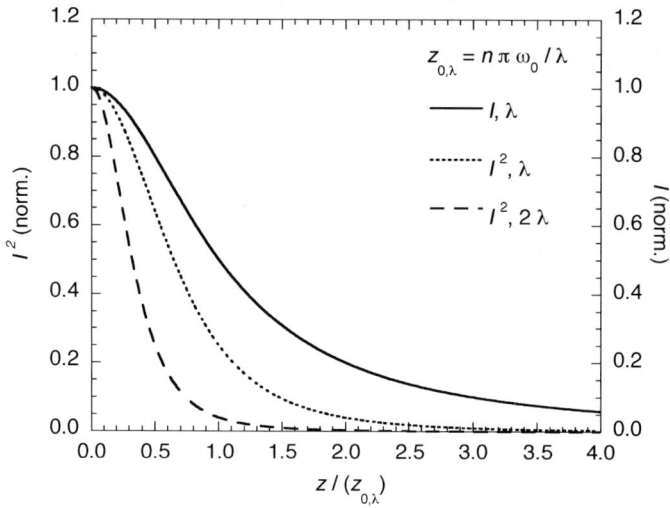

Fig. 2 Relative excitation rates for the 1PA and 2PA process along the beam propagation direction. (*Solid line*): On-axis intensity for a Gaussian beam with waist ω_0 and wavelength λ; (*dotted line*): on-axis square of the intensity a Gaussian beam with waist ω_0 and wavelength λ; (*dashed line*): on-axis square of the intensity a Gaussian beam with waist ω_0 and wavelength 2λ. In all cases, the ordinate is normalized to the value at the focus

organic molecules. Usually, pulsed lasers with high peak power are needed to study 2PA in materials and for applications based on this process.

In addition to spatial confinement, another feature of 2PA is the possibility of excitation of molecules at increased depth relative to one-photon absorption in a nominally high-absorbing medium, because the photon energy lies well below that at which the medium absorbs via 1PA. Due to these two features, 2PA has enabled the realization of 3D optical data storage [1, 2, 17, 18], lithographic microfabrication [4, 19–23], and imaging [3, 5, 24–26] via pulsed laser excitation. By computer-controlled scanning of the focus of a laser beam within a photochemically active precursor material, many complex 3D structures can be fabricated with submicron resolution using both negative- and positive-tone resists. A negative-tone resist is a type of photoresist in which the portion that is exposed to light becomes relatively insoluble in a developer, for example, due to polymerization or crosslinking, and the unexposed portion remains soluble in the developer. A positive-tone resist, in contrast, is a type of photoresist in which the portion that is exposed to light becomes more soluble in a developer, for example due to decomposition of a polymer, than the unexposed portion. Examples of structures and materials systems used for 3D lithographic microfabrication (3DLM) will be presented in Sect. 5.

3
Early Studies of Two-Photon Absorption in Organic Molecules

3.1
Introduction

Interest in the study of molecules exhibiting two-photon absorption in the visible and near-infrared range of the electromagnetic spectrum has increased significantly since the late 1990s, with the identification of a few classes of chromophores with larger cross sections than materials which were previously available (see Sect. 4). This interest was partly stimulated by the demonstration of various practical applications of two-photon absorption, such as three-dimensional fluorescence imaging, optical data storage, and microfabrication, as described above. All these applications benefited from improvements in laser technology, which made tunable ultrafast laser systems commercially available around that time. Following these discoveries, research in 2PA focused mainly on identifying organic materials with increasingly large 2PA cross sections, as these would either allow improvement of the signal-to-noise ratio or the process efficiency for a given laser intensity, or achievement of the same excitation rate in the material using lower laser intensities.

However, before the 1990s a significant number of papers on 2PA spectroscopy had already been published. Many of these are worth mentioning, not only because they laid the foundations for the field, but also because they addressed interesting spectroscopic issues for a variety of organic molecules. Indeed, at that time, 2PA was mainly used as a spectroscopic tool, in combination with the more traditional 1PA and fluorescence approaches.

The examples discussed in this section are not intended to be an exhaustive review of the early literature in the field, but should provide a taste for the range of molecular systems and experimental techniques used in two-photon spectroscopy. Often these studies involved the measurement not only of the 2PA spectrum,[4] which can be obtained, for example, by monitoring the intensity of a 2P-induced fluorescence signal as a function of wavelength (this signal being proportional to the number of molecules excited by 2PA and thus to the square of the laser intensity or the photon flux, as seen in Eq. 2), but also of the polarization ratio (Ω), defined as the ratio between the 2PA cross section for linearly and circularly polarized light. It is well known, in fact, that the probability for a 2PA transition depends on the polarization

---

[4] The term "2PA spectrum" will be used here and in the rest of the chapter to indicate the variation of the two-photon cross section as a function of wavelength, even when the data were obtained by monitoring the intensity of the fluorescence emission induced by two-photon absorption (that is when, strictly speaking, the two-photon induced fluorescence excitation spectrum was obtained). Also, unless otherwise specified, the spectra are reported as a function of the wavelength of the excitation beam and are degenerate 2PA spectra.

vectors of the two photons absorbed, even for an isotropic sample, such as a solution [16, 27, 28]. As a consequence of this, information on the symmetry of the final state in the transition can be derived from measurements of the 2PA cross section for different polarizations of the excitation beam. This is in contrast to the case of 1PA, for which the transition probabilities do not depend on the polarization state of the excitation beam, unless the sample is anisotropic.

In addition to providing a brief tutorial on 2PA spectroscopy, some of the examples discussed here can be useful in understanding the origin of the 2P activity and in estimating approximate 2PA cross sections of so-called "conventional" initiators which have been used by some investigators in microfabrication applications (see Sect. 5).

3.2
Benzene Derivatives

One of the first classes of organic compounds to be investigated in depth by two-photon spectroscopy was that of benzene and its derivatives. A review of the field prior to 1979 can be found in [29]. This early work was mainly prompted by interest in understanding the characteristics of electronic states and the way in which these affect the chemical and physical properties of the substituted benzene. In systems of high symmetry, such as benzene, the picture of the electronic structure that can be obtained by the more conventional 1PA spectroscopic approaches is definitely incomplete, as only transitions to states within a subset of the possible symmetries are formally dipole-allowed (although electronically forbidden transitions could still be observed in some cases, if sufficient oscillator strength is obtained through vibronic coupling). In addition, benzene and its derivatives have also been used as benchmarks to test theoretical predictions of spectroscopic properties at various levels of approximation.

Benzene belongs to the D_{6h} point group and 1PA transitions from the ground state (A_{1g} symmetry)[5] are dipole-allowed only to states of E_{1u} and A_{2u} symmetry [30]. 2PA transitions from the ground state are allowed to A_{1g}, E_{1g}, and E_{2g} states. Transitions to all other states are forbidden.

The lowest excited states of benzene are actually of B_{2u} and B_{1u} symmetry, and they can be reached directly by neither one-photon nor two-photon purely electronic transitions (the 0–0 band at energy E_{0-0}, the origin of the transition, is absent from the spectra). However, excitation into these states can be obtained through vibronic coupling (VC), if a vibrational mode of an appropriate symmetry is coupled to the electronic transition. The 1PA or 2PA spectra can then show a series of narrow peaks shifted from the 0–0 band

---

[5] In this contribution, only singlet states will be explicitly discussed and the multiplicity superscript of the symmetry labels of electronic states will be omitted.

by the energy of the coupling mode/modes ($E_{VC} = E_{0-0} + h\nu_\nu$, ν_ν being the frequency of the coupling mode). A Franck-Condon progression can also be present, appearing in the spectrum as a series of equally spaced bands on the high-energy side of E_{VC} (at $E_{VC} + nh\nu_{\nu'}$, where $\nu_{\nu'}$ is now the frequency of a totally symmetric mode). The 1PA and 2PA spectra of liquid benzene in the range of the B_{2u} transition (often called the L_b band, using the Platt notation [31, 32]) are shown in Fig. 3 [33].

In substituted benzenes, the symmetry is lowered and the transitions into the states that correlate to the B_{2u} and B_{1u} states of benzene become allowed by 1PA, 2PA, or both. However, when the substituents induce only a weak perturbation on the benzene π-electron system, the 1PA or 2PA spectra of the substituted compounds often closely resemble the spectrum of the unsubstituted parent molecule. Various theoretical models have been developed in an attempt to predict the type of change in the band intensity and characteristics in the 2PA spectra of substituted benzenes and, more generally, of alternant hydrocarbons [34–36]. It was found that the effect of a perturbation is quite different for 1P and 2P allowed transitions. In particular, 2P transitions to the state correlated to the benzene B_{2u} state (L_b) are affected more by vibronic coupling than transitions to the state correlated to the benzene B_{1u} state (L_a, in Platt notation [31, 32]). In contrast, inductive perturbations enhance the L_a band more than the L_b band. The effects of vibronic coupling and inductive substituents are reversed for 1P transitions into these states. Experimental

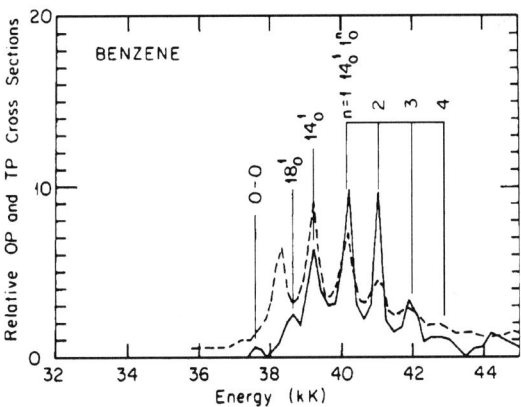

Fig. 3 1PA (*dashed line*) and 2PA (*solid line*) spectra of liquid benzene. The 2PA spectrum was obtained by the thermal lensing method. The units on the energy scale are kilokaysers (1 kK = 1000 cm^{-1}) and refer to the total energy of the transition. 0–0 is the origin of the L_b band, 18, 14, and 1 are indices for the vibrational modes (the superscript and subscript are the quantum numbers for the mode in the excited state and ground state, respectively). Reproduced with permission from [33]. © 1986, American Chemical Society

results, overall, are consistent with this description. A few examples are discussed here.

In the 2PA spectra of benzene and fluorobenzene (neat liquids) [33] the L_b band, which is observed in the range $\lambda_{exc}^{(2)} \approx 470-530$ nm (for benzene, see Fig. 3), has a clear vibronic structure with peaks at similar energies, its 0-0 component is absent or very weak, and the integrated intensity of the whole L_b band is on the same order (and is essentially due to vibronic coupling). In the 1PA spectrum, in contrast, the 0-0 band is more intense in fluorobenzene than in benzene. The 2PA spectrum of benzene in the L_a region ($\lambda_{exc}^{(2)} \approx 400-450$ nm) shows a band that is less structured and weaker than the L_b band. In contrast, the L_a band of fluorobenzene is significantly stronger than the L_b band, because of the increase in the electronic contribution to the transition probability. For phenol, the intensity of the L_a transition in the 2PA spectrum is even larger than in fluorobenzene, and both the L_b and L_a bands are red-shifted with respect to benzene and fluorobenzene [33]. Even if the L_b and L_a transitions in substituted benzenes can in some cases be identified by the appearance of weak bands in the 1PA spectrum of these compounds, only the use of a combination of linear and nonlinear spectroscopic techniques (including the measurement of the 2P polarization ratio, Ω) permitted more rigorous symmetry assignments to be made for the electronic states involved in the transitions.

Benzene has also been studied at higher excitation energies (total energy > 6.3 eV), mostly in the vapor phase or in the crystalline state at low temperatures, in an attempt to locate the valence E_{2g} state, which should be two-photon allowed and was predicted by molecular orbital theoretical approaches to be the excited state lying immediately above the B_{2u} and B_{1u} states. Experimentally, this state has not been located at energies below 7 eV. However, a resonance in the multiphoton ionization spectrum of benzene vapor indicated the presence of the 3s E_{1g} Rydberg state at 6.33 eV above the ground state (corresponding to $\lambda_{exc}^{(2)} = 392$ nm) [29, 37].

The 2PA cross section for some of the bands in benzene and its derivatives has been estimated using a variety of experimental methods. At 400 nm (in the L_a band region), the 2PA cross section of benzene was reported to be less than 0.1 GM (this value being the detection limit of the experimental method used, a two-color nonlinear transmission measurement [27]), while for the main vibrational component of the L_b band at 510 nm (sometimes referred to as the "14_0^1" band, see Fig. 3), values from 0.060 GM [38] (using a 2P-induced fluorescence method) to 8.7 GM (using a Coherent Antistokes Raman Spectroscopy, or CARS, method [39]) have been measured. Some investigators have used the strength of this transition as an internal standard in the measurement of the 2PA cross section of other substituted benzenes [40, 41], under the assumption that the intensity of this band is almost constant in this class of molecules (in light of the fact that substituents only weakly affect the L_b transition, as discussed above). However, the intensity of this band was found to be about 5 times smaller in toluene than in benzene in one study [38] and 2.5 times larger in another [33].

It should be noted that the selected results just mentioned for absolute or relative cross sections span a wide range of numerical values. The experiments required to measure these quantities are very delicate and rely on a range of assumptions and approximations that are often difficult to test on specific materials systems. In the case of small organic molecules, one additional difficulty lays in the fact that the 2PA cross sections in question are all quite small, requiring the use of very large photon fluxes to obtain data with acceptable signal-to-noise ratios. On the other hand, benzene derivatives are typically liquid at room temperature and, thus, the study of their properties can take advantage of the large number density of molecules in neat materials.

For aniline, the 2PA spectrum is significantly red-shifted with respect to benzene (the 0–0 band of L_b in the liquid is observed for $\lambda^{(2)}_{exc} = 604$ nm) and the vibronic features are not well-resolved (at least in the liquid phase) [33]. In addition, there is a large increase (about two orders of magnitude) in the intensity of the L_a and L_b bands with respect to benzene and derivatives with substituents that affect the π-system less strongly, an increase that is larger than predicted based on pseudoparity and inductive effects [33, 40]. The results are consistent with aniline having molecular π-orbitals that extend over the whole molecule and that are not confined to the benzene ring. A weak perturbation approach starting from the states of benzene is thus no longer sufficient to describe the substituted molecule.

Similarly, in benzene derivatives with unsaturated substituents, such as styrene and phenylacetylene, the π-electron systems are often delocalized over the whole molecule. To describe the substituent effect on the 2PA spectra in this case, models which include resonance contributions from charge-transfer configurations [40, 41] have been developed. Spectral features and band intensities in the 2PA spectrum of phenylacetylene vapor can be explained in the context of these models. In particular, an intense 0–0 component is observed at 558 nm, indicating that the transition is strongly electronically allowed (the 1PA spectrum at the corresponding wavelength of 279 nm is mostly due to vibronic coupling instead). The overall intensity of the L_b transition is found to be about eight times larger than in benzene (but is smaller than in aniline) [41].

Aromatic amino acids are an interesting group of substituted benzenes, because of the ever-increasing use of 2P-induced fluorescence as a microscopy tool to study systems of biological relevance. The knowledge of the 2PA properties of these amino acids could be useful either to exploit directly the auto-fluorescence of proteins or to distinguish the background fluorescence signal when an external probe is introduced in the sample. The 2PA spectra of phenylalanine and tyrosinamide, both containing a substituted benzene in their structure, together with that of tryptophan (a substituted indole) have been measured in aqueous phosphate buffer (see Fig. 4) [42]. For phenylalan-

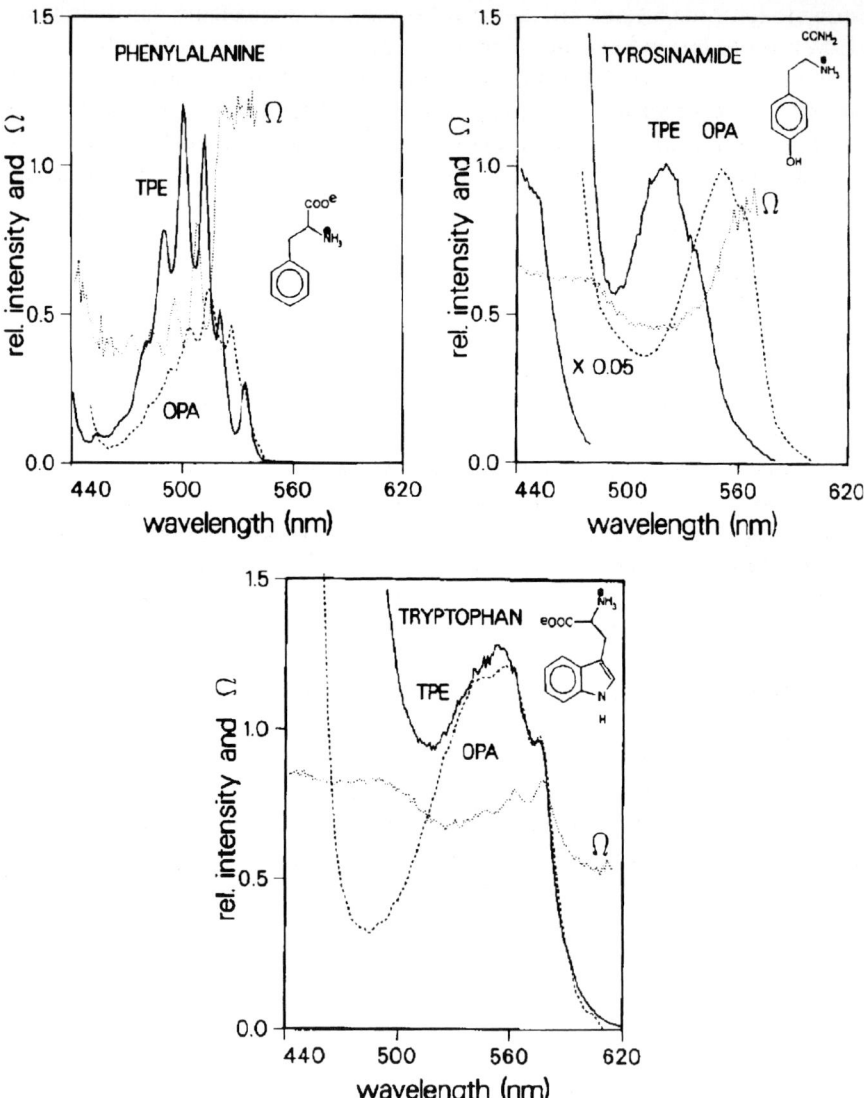

Fig. 4 1PA (*dashed line*) and 2PA (*solid line*) spectra, and two-photon polarization ratio, Ω (*light dotted line*) for: phenylalanine, tyrosinamide, tryptophan. The abscissa corresponds to the excitation wavelength of the 2PA case. The 1PA spectra are plotted at twice their excitation wavelength. The molecular structures are shown as insets in each graph. Reproduced with permission from [42]. © 1993, Elsevier

ine, both the 1PA and 2PA spectra exhibit a rich vibronic structure, which is different in the two spectra due to the coupling of a different normal mode with the electronic state in each case (in the 2PA case, the longest wavelength band is detected at 535 nm), and are similar to the spectrum of the toluene L_b transition. For tyrosinamide, no vibronic structure is seen (due to solvent interactions with the OH group) and the 2PA spectrum is blue-shifted with respect to the corresponding 1PA band, in parallel to what is observed for phenol. The 2PA cross section is approximately the same in these two compounds (only relative values of the cross sections have been reported in this study).

The situation is different for tryptophan: the 1PA and 2PA spectra almost overlap, the transition being strongly electronically allowed in both cases. The 2PA peak is located at 560 nm and the polarization ratio near the red edge of the band indicates that for this molecule the energy of the L_a transition is lower than for the L_b transition. The 2PA cross section is at least an order of magnitude larger for tryptophan than for the benzene-based amino acids. This difference in transition intensity and position could offer a way to selectively interrogate one type of amino acid or the other in a natural or synthetic protein. It should be mentioned that these spectra were obtained by a 2P-induced fluorescence method which explicitly took into account and corrected for changes in the spatial and temporal beam characteristics with excitation wavelength [43].

3.3
Larger Aromatic Molecules

Studies on the 2PA properties of organic materials have been extended from the early days to a large group of aromatic compounds beyond benzene derivatives. Here, we will focus on a few examples in which studies on series of molecules under the same experimental conditions are available and were used to gain insight into the relationship between the observed properties and the molecular structure of the systems.

3.3.1
Naphthalenes

In the case of naphthalene, transitions to the two lowest excited states (again, often indicated with L_b and L_a) are two-photon forbidden, as in benzene. However, due to vibronic coupling, the L_b band is visible in the 2PA spectrum of naphthalene in the 575–650 nm region (see Fig. 5), while L_a gains intensity in the 1PA spectrum and peaks around 275 nm [44–46], but is basically absent from the 2PA spectrum; this is again in line with predictions based on the pseudoparity of the states. Polarization ratio data were used to aid the band assignment. A weak 0–0 peak of the L_b band can actually be seen in the 2PA spectrum (at 630.5 nm for naphthalene in cyclohexane [45] and at 631.8 nm in carbon tetrachloride [47]), probably because of local perturbation of the symmetry due to the solvent environment or other effects [44, 45]. The 2PA

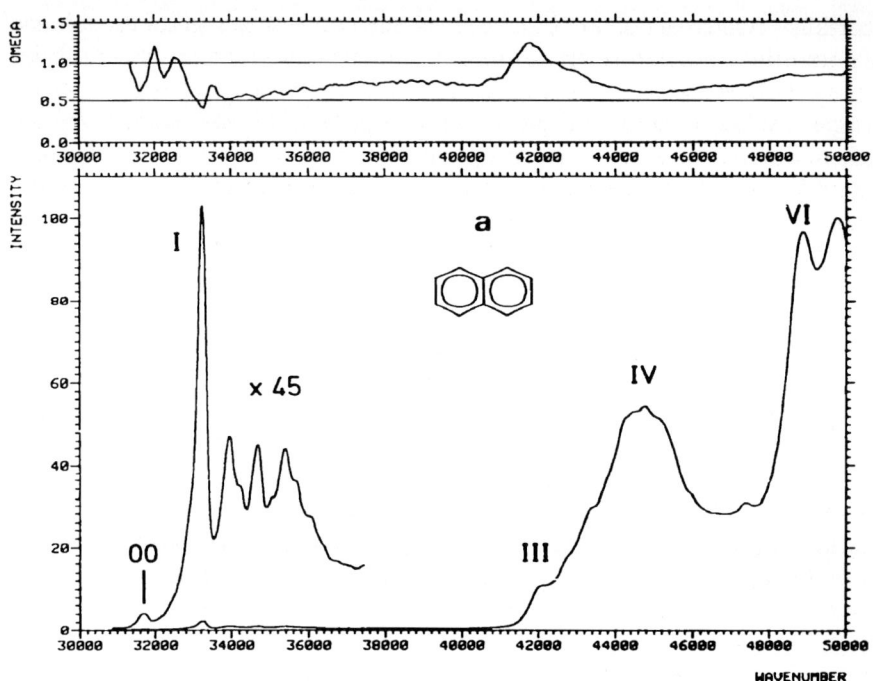

Fig. 5 *Bottom*: 2PA spectrum for linearly polarized light (intensity in arbitrary units) and (*top*) polarization ratio of naphthalene in ethanol. 0–0 is the origin of the L_b band (I), the L_a band (II) is not visible, and the numbers III–VI identify transitions to higher electronic states. The abscissa, representing the total excitation energy, is in units of cm^{-1}. Reproduced with permission from [46]. © 1981, Elsevier

cross section is at least an order of magnitude larger below 500 nm than in the L_b band region and a broad peak can be seen at 449–451 nm, corresponding to a transition to an excited state with *gerade* symmetry, which is two-photon allowed [45, 46].

It should be pointed out, however, that the details of the vibronic structure of the L_b band are significantly different in some of the spectra which have appeared in the literature and which have been mentioned above. For example, while Jones et al. [44] and Dick et al. [46] have reported only one relatively strong peak in the range 600–625 nm, Mikami et al. [45] observed two peaks. There are also differences in the relative intensities of the L_b transition and the transition to the *gerade* state around 450 nm. In [44, 45], the spectra were obtained in the same solvent (cyclohexane) and for similar concentrations (0.05 M in one case, 0.1 M in the other), while in [46] the solvent used was ethanol. In all three cases the experiments utilized linearly polarized light and detected the 2P-induced fluorescence. The observed differences are probably ascribable to some other experimental condition or a source of error that was not accounted for. This is another example of the reproducibility issues encountered in the measurements of 2PA spectra in different laboratories or using different techniques, which, to some extent, still exist today. This problem can affect the interpretation of the spectra, especially in cases where data on the fine structure of 2PA bands are analyzed in detail to obtain, for example, information on which of the vibrational modes provide intensity to the band by vibronic coupling or appear in a progression.

The effect of substitutions on the 2PA spectrum of naphthalene is similar to that in benzene. For example, chloro-substitution and fluoro-substitution at the 1-position and, more so, at the 2-position, leads to an increase in the intensity in the L_a band in the 2PA spectrum (and, conversely, to an increase of the L_b intensity in the 1PA spectrum), while the L_b transition changes little in shape and intensity. Overall, both bands are still weak with respect to transitions to higher-lying states related to totally symmetric states of naphthalene, even though the L_a and L_b 2PA transitions are dipole-allowed in these substituted naphthalenes [44]. In the case of isoquinoline, in which the carbon atom in position 2 is replaced by a nitrogen atom, the effect of the inductive substituent is stronger and the L_a and L_b bands have comparable intensity in both the 1PA and 2PA spectra. However, the 0–0 component of the L_b band is still very weak in the 2PA case, indicating that most of the intensity to this band derives from vibronic coupling [44].

3.3.2
Indoles and Derivatives

In indole, benzimidazole, and their derivatives, all symmetry restrictions on the 1PA and 2PA allowedness of transitions are lifted. In these molecules, L_a and L_b transitions are not formally defined. However, bands in the low-energy portion of the 1PA and 2PA spectra can be assigned to transitions which have similar characteristics to the L_a and L_b transitions of benzene (although they are red-shifted with respect to benzene or toluene), so the nomenclature is often extended to these systems.

For indole and methylindoles, the 0–0 component of the L_b-type band is quite strong in the 2PA as well as in the 1PA spectrum, indicating that the intensity of the band has a significant electronic contribution [48]. Even if there are some differences in the relative intensities of some of the vibronic components of the L_a- and L_b-type transitions, the 1PA and 2PA spectra look quite similar for each compound (when represented as a function of the total transition energy) [48]. For these molecules, then, vibrational modes with similar frequencies couple with a given electronic state and give rise to the observed vibronic structure of the corresponding 1PA and 2PA bands. This may not happen in other classes of chromophores. The assignment of the observed peaks in the 2PA spectrum of these compounds is complicated by the fact that the L_a- and L_b-type transitions are affected differently by the solvent environment, and the order of the corresponding electronic states can change on switching between polar and nonpolar solvents [48, 49].

The 2PA spectrum of benzimidazole is blue-shifted significantly with respect to indole, the 0–0 band appearing at 569 nm for benzimidazole in isopropanol [50] and at 591 nm for indole in butanol [48] (and at 585 in hexane [50]). It should also be mentioned that the intensity of the 2PA bands for benzimidazole is only slightly stronger than that for toluene, while the peak 2PA cross section for indole is at least five times larger than for toluene (although cross section values were not measured in this study) [50]. This is in agreement with the observation, mentioned above (see Sect. 3.2), that tryptophan, which is a substituted derivative of the indole π-system, has much larger 2PA cross section than the other aromatic amino acids discussed.

3.3.3
Biphenyls

A series of biphenyls has also been investigated by 2P-induced fluorescence. This particular experiment was designed in such a way that the polarization ratio could be obtained from the simultaneous measurement of the fluorescence signal for linearly and circularly polarized light [51, 52]. Two bands can be observed in the lowest energy portion of the 2PA spectrum of biphenyl (in carbon tetrachloride), one peaking at 565 nm, the other at 536 nm. It is interesting to note that both bands correspond to transitions into states at lower energy than the main band in the 1PA spectrum (≈ 250 nm). The polarization ratio data also indicate that these states have different symmetry from one another, the lowest energy state belonging to one of the B species, the other being possibly of A-type. The electronic origin of the B-type state can be seen (at 604 nm) in the 2PA spectrum of biphenyl crystals, as expected for an allowed transition [53]. A much stronger 2PA peak than the two mentioned above is observed at 445 nm in solution [51]. In 4,4′-dichloro- and 4,4′-difluoro-biphenyl, the same two low-energy bands can still be seen in the 2PA spectrum and have approximately the same relative intensity as for biphenyl. This is because the inductive effect of the substituents is small and the twist angle between the rings is similar to that in biphenyl (biphenyl is not planar in solution). Steric hindrance is probably responsible for a decrease, with respect to biphenyl, in the intensity of the transition into the B band for 2,2′-difluorobiphenyl, for which the twist angle is much larger.

In biphenyls bridged at the 2 and 2′ positions, the inversion center of the ideal planar biphenyl is removed and transitions can be allowed in both 1PA and 2PA cases. As a consequence, the 1PA and 2PA peaks are observed at approximately the same energy for fluorene (the lowest 2PA peak is located at 586 nm) [52]. Similar observations hold true for carbazole, dibenzofuran, and dibenzothiophene. However, the fine details of the spectra are hard to interpret. In a number of cases, 2PA peaks actually appear at the edge of the tuning range of a given laser dye used to operate the laser. Therefore, doubt can arise suggesting that some of the spectral features may be artifacts of the experimental method (due to variations in the second-order coherence or spatial profile of the laser) and not actual 2PA peaks.

Even if all transitions are formally allowed in these noncentrosymmetic molecules, their relative intensity in the 1PA and 2PA spectra can be different. In fact, in the case of alternant hydrocarbons, transitions to a state of "–" pseudoparity are weak in the 1PA spectrum and strong in the 2PA spectrum, while transitions to a "+" state are strong in the 1PA and weak in the 2PA spectrum [40]. This can easily be seen in the spectrum of phenanthrene, with a relatively strong 2PA band at 658 nm (A_1 state) and a weak one at 587 nm (B_2 state) [54]. Conversely, the transition to the B_2 state is more intense than that to A_1 in the 1PA spectrum. Higher-energy 2PA states with large 2PA cross sections can also be seen for this compound in the range $\lambda_{exc}^{(2)} = 440$–530 nm .

3.4
Polyenes

A class of materials that received considerable attention in the early days of this research field, and that still remains of interest, is that of polyenes. Studying the properties of polyenes was and is relevant, for example, because they can be considered as prototypes for many organic semiconductors and nonlinear optical materials, and because derivatives of the polyene retinal are involved in the vision process in biological systems. Their characteristic properties are due to the highly polarizable π-electrons in the conjugated backbone of these chromophores. Results of spectroscopic investigations indicated that the ordering of the excited states of most linear polyenes, except the shortest elements of the series, is different from that predicted by simple molecular orbital theory at the Hartree-Fock level [55, 56]. In particular, large gaps between absorption and fluorescence bands, different effects of solvent polarity on absorption and fluorescence spectra, low fluorescence quantum yields, long fluorescence lifetimes, and dual fluorescence emission, among other evidence obtained by researchers, suggested that in many polyenes the lowest-lying electronic state is not a strongly 1P-allowed B_u state, but a 1P-forbidden A_g state. If the effect of electron correlation is taken into account in quantum chemical calculations by the inclusion of double-excitation configuration interaction, an A_g state is indeed predicted to be at lower energy than the first B_u state in butadiene and longer polyenes [57].

It was mentioned above (see Sect. 2) that for centrosymmetric molecules, such as *all-trans* polyenes and α,ω-disubstituted *trans*-polyenes, there is a complementarity between the selection rules for 1P- and 2P-allowed transitions. Transitions from the ground state (which is of *gerade* type) to a B_u state are visible in the 1PA spectrum, while transitions from the ground state to an A_g state appear in the 2PA spectrum. For this reason, 2P spectroscopy has been the tool of choice to investigate the nature of the low-lying A_g electronic state of polyenes. Some of the findings are discussed below.

In the case of *trans*-stilbene, it was found that the 2PA spectrum in methylcyclohexane is relatively broad and unstructured, with a maximum at 476–500 nm [58]. The value of the two-photon polarization ratio Ω is consistent with an A_g state and with a transition tensor that lies in the plane of the double bond. The lowest 2P electronic state is thus located between those that give rise to the two main bands in the 1PA spectrum of this compound in the ultraviolet range (294 and 229 nm). The 2PA cross section of stilbene in cyclohexane has been measured by a three-wave mixing technique at 514.5 nm and found to be 12.1 ± 0.9 GM [59, 60]. From this value of the cross section and the shape of the 2PA spectrum reported by Stachelek et al. [58], the 2PA cross section at the maximum of the lowest 2P band, δ_{max}, of stilbene can be estimated to be around 18 GM. However, the estimate for δ_{max} would be a significantly larger number (≈ 60 GM) if the stilbene 2PA spectrum were scaled using the 2PA cross section[6] measured at 532 nm by an absolute 2P-induced fluorescence method implemented by Chen et al. [61].

The 2PA spectrum of *E,E*-1,4-diphenyl-1,3-butadiene has been measured by 2P induced fluorescence in the range 670–730 nm (in cyclohexane) [62] and by thermal lensing in the range 570–660 nm (in chloroform) [47]. In this case, a feature observed at 709 nm in the 2PA spectrum is assigned to an A_g state that is slightly lower in energy than the B_u state (the lowest-energy state in the 1PA spectrum, which is observed at 347 nm). The 2PA cross section is estimated to be about 1.0 GM at 707.7 nm and increases at shorter excitation wavelengths, but no clear peaks are visible in the spectrum down to 570 nm [47, 62]. Based on the cross section at 707.7 nm and the spectral shape obtained by splicing the data reported in Refs. [62] and [47], the cross section at 600 nm could be estimated to be larger than 100 GM. However, the 2PA cross sections actually measured are significantly lower: 30 GM at 600 nm [47] and 14.4 ± 2.0 GM at 608 nm [60].

As briefly mentioned in Sect. 3.2, the disagreement between cross section values obtained by different experimental techniques and different research groups is not unusual and it is mainly ascribable to the difficulty in accounting for all the sources of error in the experiments and in assessing the applicability of underlying assumptions. For example, it is usually very difficult to describe the spatial and temporal characteristics of the laser beam and how these vary with wavelength. For this reason, it is often assumed that either they do not change with wavelength or that they can be described by simple analytical expressions [63–65]. This approximation can lead to errors in single wavelength measurements of cross section or to artifacts in spectral

---

[6] Due to a difference in the definition of the cross section in the work by Chen et al. [61], the value of 8 GM that appears in the paper actually corresponds to $\delta = 16$ GM in the conventions utilized in the present contribution, which follow Eq. 2.

features [43]. However, all the measurements for diphenylbutadiene indicate that the observed 2PA transitions are to electronically allowed states.

In a separate study, the cross section of diphenylbutadiene in chloroform has been measured at 532 nm by two different methods and reported to be 40 ± 8 GM by degenerate four-wave mixing, and 34 ± 12 GM by nonlinear transmission [66]. It should be pointed out that both results were obtained using ns excitation pulses, but that the cross sections obtained by the two methods are comparable one to the other and on the same order of the cross sections listed earlier (although those were obtained in a different wavelength range).

Longer all-*trans* α, ω-diphenylpolyenes also exhibit an A_g state close to or below the 1P-allowed B_u state: the lowest 2P state is observed in cyclohexane solutions at 769 nm for diphenylhexatriene ($n = 3$) and around 885 nm in diphenyloctatetraene ($n = 4$) [67]. In both cases, the 2PA spectra in solution do not exhibit a strong and sharp origin and are relatively featureless, with cross sections that increase towards shorter excitation wavelengths. However, measurements at low temperature and in solid matrices allowed the assignment of the origin of the $2A_g$ state to the peak observed at 798 nm for $n = 3$ [68], 990 cm^{-1} below the origin of the $1B_u$ state. Various vibronic components of the same transition can also be resolved in the low temperature spectrum. Similarly, a progression of vibronic bands is present in the low temperature spectrum of diphenyloctatetraene, the $2A_g$ origin in this case being at 894 nm (about 2000 cm^{-1} below the origin of $1B_u$) [69]. Typically, the energy difference between the $2A_g$ and $1B_u$ state increases with increasing chain length. The cross sections at 608 nm are 43.3 ± 4.0 and 61.0 ± 23.0 GM for the $n = 3$ and 4 cases, respectively [60]. Overall, an increase in cross section with the length of the conjugated bridge is observed in this class of compounds. However, because the measurements were performed at a single wavelength and this was located outside the region where the spectral shape had been satisfactorily investigated, it is not possible to obtain quantitative structure/property relationships and to estimate the cross section for the lowest lying state from these data.

In the case of all-*trans*-1,3,5,7-octatetraene, the confirmation that the lowest excited state is of A_g symmetry was first obtained by measuring the 1PA and 2PA spectra of the compound at very low temperature (4.2 K) in *n*-octane [70]. As no local asymmetries or distortions are introduced by the host material, the full symmetry of the polyene is preserved in this experiment. It was then possible to determine the position of the origin (0–0 band) for the lowest transition in the 2PA spectrum, observed for this compound at 700.2 nm. No features were observed at the corresponding wavelength in the 1PA spectrum, while a band was observed at 348.9 nm. This band corresponds to a transition to the same state, but is shifted (by about 100 cm^{-1}) because of vibronic coupling. As a consequence of the complete absence of the origin of the transition in the 1PA case, it is possible to conclude that the final state is of A_g type. From measurements at higher temperature or in hosts in which the molecules could be located in asymmetric or disordered sites, it would not be possible to make state assignments conclusively, either because of the broadness of the bands or because the asymmetry renders the transition allowed. A wealth of other information on the electronic states involved in the transition can actually be obtained from low-temperature 2PA spectra. An illustrative example is displayed in Fig. 6 for the same molecule [71]. In the wavelength range included in the figure, corresponding to $\lambda_{exc}^{(2)} = 490–703$ nm, a multitude of narrow peaks of various intensities can be seen. The origin of the transition occurs at 700.2 nm, as discussed above. All the other peaks (72 in total), however, have been assigned to vibronic components of the same transition. It is worth noticing, in particular, that various progressions are visible, for example for the vibrational mode with frequency of about 1750 cm^{-1} (at least six overtones of this mode can be identified). This means that only one 2P-allowed excited state exists in this wavelength range. The next excited state has been identified at 447 nm (not shown in the figure), that is 2.0 eV above the 2A_g state (higher lying A_g states are sometimes generically

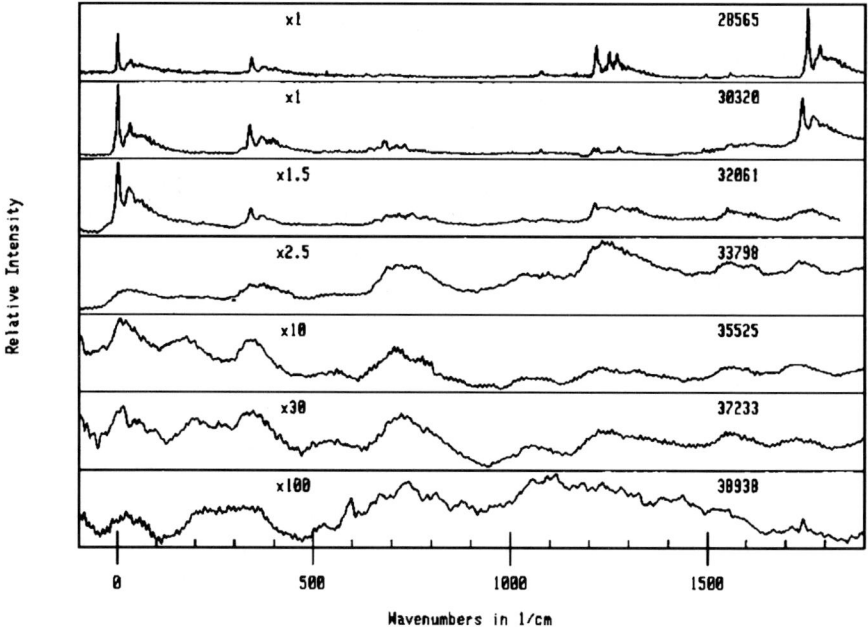

Fig. 6 2PA spectrum of octatetraene in *n*-octane at 1.8 K. The total excitation energy is given by the sum of the number in the upper right corner of each strip and the values in the abscissa (both are in cm^{-1}). There is an overlap of about 300 cm^{-1} between two consecutive strips. The numbers after "x" in each panel are magnification factors for the ordinate. Reproduced with permission from [71]. © 1996, American Institute of Physics

called mA_g states in the literature). The band for the 2PA transition to this state in octatetraene is relatively broad even at 1.8 K and its intensity 2.3 times that of the $2A_g$ origin [71].

Finally, it should be mentioned that the vibrational modes responsible for the sub-structure in the 2PA spectrum are those for the final state of the transition, that is the excited state $2A_g$ in the case of Fig. 6. It has been noticed for this compound and other polyenes, that the frequency for the mode typically assigned to the symmetric C=C stretching is higher in the excited state than in the ground state (for octatetraene, the frequencies are 1753 and 1608 cm^{-1}, respectively) [72]. This trend is not predicted by molecular orbital theory at the Hartree-Fock level. Various explanations have been proposed for the experimental observation, including vibronic coupling between ground and excited state, an increase in the C=C bond order in the excited state with respect to the ground state, and the mixing of other modes with the C=C stretching in the two states.

3.5
Xanthenes and Other Laser Dyes

Xanthenes and other deeply colored highly fluorescent dyes have been extensively used in tunable dye lasers to provide emission with wavelength spanning most of the visible and the near-infrared range. Despite the easy access to this class of materials and the number of papers relating to some aspects of their 2PA properties, the nonlinear spectroscopy of these chromophores has not been fully explored. Indeed, most of the reports are limited to the measurement of the 2PA cross section at one wavelength (typically 532 nm, 694 nm, or 1.06 μm) for one compound or a limited number of compounds. In many cases, these molecules were systems of choice to test new experimental setups for the measurement of 2PA cross sections. The level of agreement between the results from independent investigation is not always satisfactory, but it is hard, in general, to assess whether the differences are due to experimental uncertainties and systematic errors, or due to variations of the material properties with solvent, concentration, temperature, etc. A few investigators, however, have reported 2PA spectra of selected compounds (mainly rhodamines, fluoresceins, and coumarins) over a range of excitation wavelengths sufficiently wide to identify transitions to one or two electronic excited states. A complete review of the literature on these compounds is outside the scope of this contribution. We will focus here on a few interesting examples.[7,8]

---

[7] Many of the studies discussed here contain critical reviews of the prior literature on specific compounds and the interested reader is referred to them.

[8] A few other compounds for which 2PA studies appeared in the literature are: coumarin 480, 450, and 6H in [61]; acridine in [73]; acridine red in [74]; rhodamine 560 and 640, oxazine 725 in [75]; and malachite green in [76].

Many of the rhodamines and a few other laser dyes have been shown to have relatively large 2PA cross sections at 1064 nm. For example, the cross sections for rhodamine B, rhodamine 6G, and DODCI have been reported as: 14, 5.5, and 22 GM, respectively, by Bradley et al. [74], and as: 12, 11, and 38 GM, respectively, by Li et al. [75][9] If the uncertainties reported by the authors are taken into account, the two sets of data are in reasonable agreement (although different solvents were used in the two studies). For rhodamine B and 6G, for the 1064 nm excitation wavelength, the final state of the transition is one of the vibrational levels of the lowest excited state, which is two-photon allowed for chromophores lacking an inversion center. To a first approximation, the 2PA cross section for this transition depends on the change in dipole moment between ground and excited state and the transition dipole moment between these two states (see Eq. 5 below). The latter is large for rhodamines, so even a small change in dipole moments would lead to a significant cross section. The magnitude of the cross section and the high fluorescence quantum yield partly explain the extensive use of rhodamine derivatives in 2P fluorescence imaging applications.

The cross sections reported above for rhodamine B and 6G are also in good agreement with the measurements by Hermann et al. [77], which were performed using an absolute 2P-induced fluorescence method which is independent of changes in the spatial and temporal coherence of the laser. These authors have also shown that the cross section for rhodamine B and 6G is much larger at 694 nm than at 1064 nm. Based on the polarization ratio at 694 nm ($\approx 3/2$), this transition has been assigned to an A_1 state. An additional excited state, intermediate in energy between the two discussed above, has actually been identified and it gives rise to a peak in the 2PA spectra around 800–820 nm [76, 78, 79]. Most of these compounds can thus be used for 2PA applications in a relatively large range of excitation wavelengths in the near-infrared region (specifically in the typical tuning range of Ti:sapphire femtosecond lasers).

In xanthenes, even if all one-photon allowed transitions are also two-photon allowed, the shape of the bands and their relative intensities are very different in the 1PA and 2PA spectra [76, 78]. This is not the case for other laser dyes and chromophores, for which the two spectra are almost identical (if represented as a function of the total transition energy), showing peaks in the same position and with very similar band shapes. Some example of chromophores in this category are: coumarin 307 [78], coumarin 102 [80], 7-hydroxycoumarin [81], lucifer yellow [78], and cascade blue [78].

It is generally assumed that the fluorescence emission spectrum of a compound is the same after the compound undergoes 1PA or 2PA excitation. This has been confirmed experimentally in a number of cases [74, 78], but there

---

[9] These are the 2PA cross section values as reported in the two original papers. Due to a difference in the definition of δ used by those authors, the numerical values should be multiplied by two if the definition of Eq. 2 is used instead.

are few instances where the situation is different. For example, the emission spectrum of 7-hydroxycoumarin exhibits two distinct bands in the case of 1PA, but only one of them appears in the 2PA-induced fluorescence spectrum [82]. This could introduce additional difficulties in the measurement of 2PA cross sections [81], and is a reminder that the shape of the emission spectrum should always be checked.

Because of the fact that laser dyes are highly fluorescent, are often thermally and photochemically stable, and many exhibit significant 2PA cross section, as discussed above, they have been used to test the dependence of 2P-induced fluorescence signals on the laser intensity[10] over wide ranges of intensities and to identify processes responsible for deviations from the expected quadratic dependence. In many cases, the dependence becomes subquadratic above a critical intensity. In cases where saturation of 2PA absorption can be excluded and if only a small fraction of the incident photons are absorbed by the material (such that the pulse energy effectively does not change in the sample), the subquadratic behavior can be due to stimulated emission, which reduces the fluorescence intensity [77, 83, 84]. The effect is more significant the closer the excitation wavelength is to the peak of the fluorescence spectrum, because the stimulated emission cross section is proportional to the ratio $F(\lambda_{exc}^{(2)})/F(\lambda_{max,f})$ (where $F(\lambda)$ is the fluorescence intensity at wavelength λ, $\lambda_{exc}^{(2)}$ the wavelength of the excitation beam, and $\lambda_{max,f}$ the wavelength of the fluorescence maximum). A subquadratic behavior at high photon flux can also be due to excited state absorption (taking place before the molecules relax to the ground state), if the relaxation process from the final state to the state e is less efficient than from the 2P-allowed state, because of the existence of other deexcitation pathways or because the molecule undergoes dissociation [84]. Finally, a subquadratic dependence on laser flux can also be observed at low fluxes if the 1PA cross section at $\lambda_{exc}^{(2)}$ is not negligible [77].

3.6
Absolute Two-Photon Absorption Cross Sections and Spectra

Absolute measurements of the 2PA cross section of chromophores are experimentally very demanding and difficult to implement for routine studies of new materials. This is mainly due to the fact that the 2PA excitation rate in a sample does not depend simply on the *square* of the *average* intensity, $\langle I(r, t) \rangle^2$, but on the *average* of the *square* of the intensity, $\langle I(r, t)^2 \rangle$ (r and t are the special and temporal coordinate) [63, 64, 78]. These two quantities are not the same if the laser pulse has substructure in space and time [85, 86]. Thus, absolute 2PA determinations that rely on the measurements of signals proportional to the excitation rate, such as the 2P-induced fluorescence

---

[10] More precisely, the quantity of interest is the photon flux, or the number of photons arriving at the sample per unit area and time. The photon energy is the proportionality factor between the two quantities.

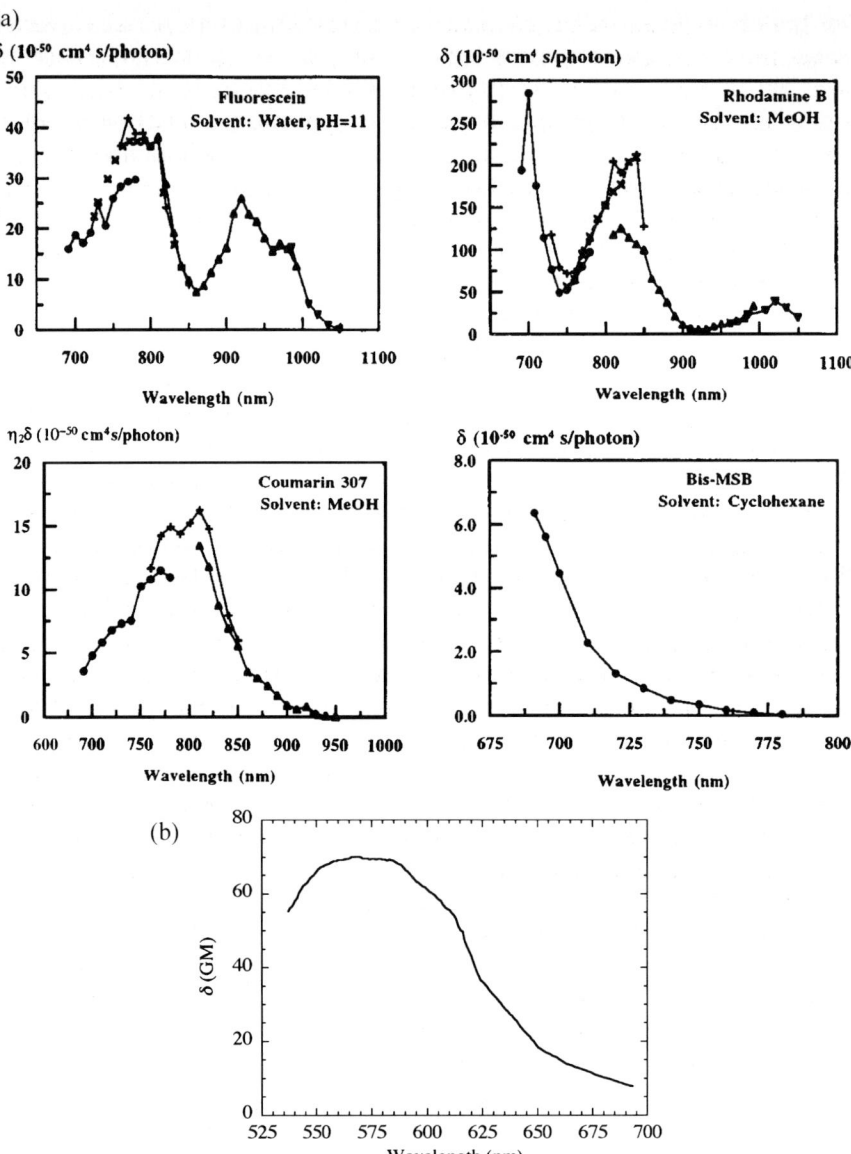

Fig. 7 Selected 2PA spectra obtained by absolute fluorescence-based methods. **a** Spectra for fluorescein, rhodamine B, coumarin 307, and *E,E-p*-bis(*o*-methylstyryl)benzene (the solvent is indicated in the legend) obtained by Xu and Webb [78]. In the case of coumarin 307, the ordinate displays the quantity $\eta\delta$, where η is the fluorescence quantum yield. **b** Spectrum for *E,E-p*-bis(*o*-methylstyryl)benzene (this spectrum is obtained from the tabulated values for the band shape reported by Kennedy and Lytle [90] and the cross section at 585 nm reported by Fisher et al. [80], to correct for a typographical error in the 1986 paper). Part (**a**) reproduced with permission from [78]. © 1996, Optical Society of America

method, require an independent measurement of the $\langle I(r, t)^2 \rangle$ quantity or, equivalently, of the second-order coherence factor, $g^{(2)}$ [64, 78, 87]. Due to the difficulty in characterizing these experimental parameters, for many of the examples discussed in Sects. 3.3–3.5, 2PA cross sections were determined for few selected wavelengths, or spectra were obtained assuming that $g^{(2)}$ was independent of wavelength (and, in some cases, it was assumed to be equal to 1). Experimental techniques based on the measurement of the beam attenuation due to 2PA, instead of the excitation rate, may not depend directly on the $g^{(2)}$ factor, but still require an accurate characterization of the pulse spatial and temporal characteristics [88, 89].

Nonetheless, absolute 2PA spectra have been reported for a limited number of chromophores. Figure 7 displays 2PA spectra measured by an absolute 2P-induced fluorescence (2PIF) technique for four compounds: fluorescein, coumarin 307, rhodamine B, and E,E-p-bis(o-methylstyryl)benzene [78, 80, 90]. These materials are commercially available, have relatively broad 2PA bands and sizable peak cross sections, and have been widely used in recent years as internal standards in 2PA spectroscopy. It should be mentioned that the uncertainty for the cross section obtained by Xu and Webb were reported to be between 25% and 33% [78]. Kennedy and Lytle did not report the uncertainty in their result for δ [90]. Also, some discontinuities can be seen in the spectra of Fig. 7a, usually in connection with wavelengths at which different mirror sets were needed to tune the laser. The field of 2PA spectroscopy could benefit from the availability of absolute 2PA cross sections for a wider range of chromophores or of excitation wavelengths and from an improvement in the precision of the absolute values. Nonetheless, the availability of the data in Fig. 7, and wavelength-dependent 2PA cross section values for a few other compounds published in the late 1990s, has facilitated the development of the field, effectively providing reference materials and data that can be used for relative measurements of 2PA cross section, which can be implemented more readily than absolute measurements. Although efforts to improve measurement techniques and test new detection schemes have continued to the present day, subsequent research in the field of 2PA has mainly focused on identifying and studying chromophores with increasingly large cross sections, as discussed in the next section.

4
Structure/Property Relationships

4.1
Introduction

All the practical applications of 2PA mentioned in Sect. 1 would benefit from the availability of compounds with large 2PA cross section, because this will either reduce the exposure time or the photon flux required to obtain a certain change in the material or collect a signal of a given magnitude, with the excitation rate for the 2PA process being proportional to the cross section (see Eq. 2). As such, intensive research efforts have been devoted to the identification of molecules with large 2PA cross sections and of general "rules" for the design of such molecules. Obviously, a candidate material for a specific application cannot be chosen solely based on the magnitude of the 2PA cross section at the wavelength of interest, because it also needs to be characterized by a high efficiency for the desired chemical or physical process that takes place after the excitation (such as fluorescence emission or chemical reactivity). Some of these additional requirements and the way they were included in the design of 2PA chromophores will be discussed in Sects. 5.2 and 5.3.

Based on experimental and theoretical results on model compounds and interpretative models derived from simplified sum-over-state expressions, it has been possible to identify specific relationships between the magnitude of the 2PA cross section and the molecular structure of the chromophore. This knowledge is invaluable in guiding the search for subsequent generations of materials with improved performance. The number of molecules specifically synthesized following these initial guidelines and investigated for 2PA applications continues to increase. In the sections below, we will discuss the main findings for various classes of chromophores.

4.2
Quadrupolar Dyes

The class of compounds that has been most extensively investigated from the point of view of two-photon absorption is that of so-called *quadrupolar* chromophores. In essence, these molecules are linear conjugated chains with electron donating or withdrawing substituents arranged symmetrically with respect to the center of the molecule (Fig. 8, classes I–IV). With the inversion center being preserved, the lowest order moment supported by these molecules is the quadrupole moment.

Restricting the sum-over-states equations, which in general cases need to include all states of the system [91, 92], to include only terms relating to three states of a quadrupolar molecule, the ground state (g), the lowest one-photon allowed excited state (e), and the lowest two-photon-allowed state (f), which are the states depicted in Fig. 1 when $f = e'$, the maximum two-photon cross section can be expressed as follows [93, 94]:

$$\delta_{max} = \frac{\hbar\omega^2 L^4}{5\epsilon_0^2 c^2 n^2} \frac{M_{ge}^2 M_{ef}^2}{(E_{ge} - \hbar\omega)^2 \Gamma} \tag{3}$$

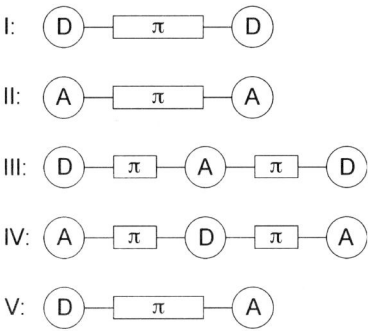

Fig. 8 Schematics of various linear (one-dimensional) chromophores classified based on the substitution pattern. I–IV: quadrupolar molecules; V: dipolar molecules (D = donor group; π = π-conjugated bridge; A = acceptor group)

(with all quantities expressed in S.I. units) or:

$$\delta_{max} = \frac{16\pi^2 \hbar \omega^2 L^4}{5c^2 n^2} \frac{M_{ge}^2 M_{ef}^2}{(E_{ge} - \hbar\omega)^2 \Gamma},$$

(4)

(in cgs units). In Eqs. 3,4, M_{ge} and M_{ef} are, respectively, the transition dipole moments between the states g and e and between e and f. E_{ge} is the energy of state e with respect to g, $\hbar\omega$ is the energy of the excitation photon ($\lambda_{max}^{(2)}$ is the corresponding wavelength) at the 2PA resonance (i.e., $2\hbar\omega = 2hc/\lambda_{max}^{(2)} = E_{gf}$, with E_{gf} equal to the energy of state f with respect to g), and Γ is an energy damping term for the transition $g \to f$. The quantity $E_{ge} - \hbar\omega$ is often called detuning energy, ΔE.

It is evident from Eqs. 3,4 that δ_{max} should be largest for compounds characterized by large values of M_{ge} and/or M_{ef}. The cross section also increases when the detuning energy decreases. One strategy to increase the magnitude of δ_{max} is to build molecules with electron-rich groups at the termini of the conjugated bridge (as in class I of Fig. 8). It was shown that this substitution affects mostly the transition moment M_{ef}. For example, for *trans*-stilbene, **q.1**, $\delta_{max} \approx 18$ GM[11] and for E-4,4'-bis(dibutylamino)stilbene (**q.2** in Table 1) $\delta_{max} \approx 2.0 \times 10^2$ GM[12] [95]; that is, the bis-donor substitution leads to an increase in cross section of an order of magnitude. The values of M_{ef} calculated from an INDO-MRD-CI approach are 3.1 and 7.2 D for **q.1** and **q.2**, respectively [95]. Quantum chemical calculations also showed that the charge distribution in **q.2** is significantly different in states g and e, the transition

---

[11] Value obtained using the result at 514.5 nm by Anderson et al. [59] to scale the spectrum reported by Stachelek [58] (see Sect. 3.4).

[12] In view of the typical experimental uncertainty in the measurement of 2PA cross sections (10–20%), we choose in this section to report only two significant digits for all δ values and not to include the errors in the measurements, even when specified in the original work.

Table 1 2PA data for class I of linear choromophores with diphenylpolyene backbones (solvent: toluene)

	Molecule	$\lambda^{(2)}_{max}$ (nm)	δ_{max} (GM)	Refs.
q.2		600	2.0×10^2	[94, 95]
q.6		640	2.6×10^2	[94]
q.7		710	3.2×10^2	[94]
q.8		730	4.3×10^2	[94]
q.9		730	1.3×10^3	[94]

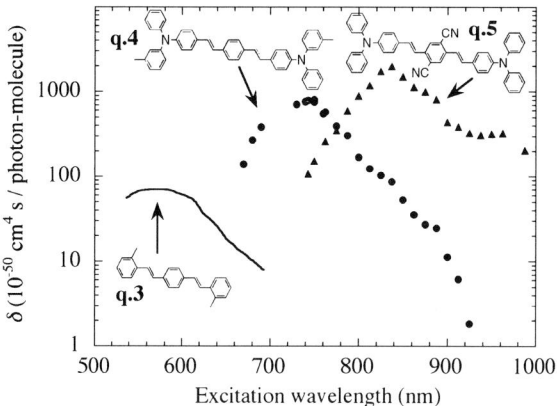

Fig. 9 Two photon absorption spectra (from 2PIF measurements) of distyrylbenzene derivatives with various substitution patterns. **q.3**: data from [80, 90]; **q.4** from [94]; **q.5** from [95, 96]

$g \rightarrow e$ being characterized by a charge transfer from the amino groups toward the central double bond. Structural modifications and substitutions that facilitate this charge transfer and increase the change in quadrupole moment are expected to result in an increase in δ_{max} [95].

A similar trend can be observed in quadrupolar molecules with a distyrylbenzene bridge, as shown in Fig. 9. In the molecule without π-donor or acceptor substituents, **q.3**, $\delta_{max} = 70$ GM at 568 nm,[13] while for chromophore **q.4** (belonging to class I in Fig. 8) the 2PA band shifts to longer wavelength, 745 nm, and increases in magnitude ($\delta_{max} = 8.0 \times 10^2$ GM). When electron withdrawing groups are present in the center of the π-bridge, creating a D–A–D pattern (as in class III of Fig. 8), the cross section is even larger, with $\delta_{max} = 1.9 \times 10^3$ GM at 835 nm for **q.5** [95]. Once again, the increase is mostly due to a change in the transition moment M_{ef}, which increases in the order: 2.8 D, 12 D [94], and 15.5 D [96] for **q.3**, **q.4**, and **q.5**.[14]

The 2PA cross section also increases when the π-conjugated bridge is extended, as shown for D-π-D α, ω-diphenylpolyenes **q.2**, **q.6**–**q.9** in Table 1: the value of δ_{max} increases from 2.0×10^2 to 1.3×10^3 GM going from 1 to 5 double bonds in the chain [94]. A red shift in the position of the 2PA peak accompanies the increase in cross section. A similar trend in $\lambda_{max}^{(2)}$ and δ_{max} is observed when the chain length is increased from **q.2** to **q.10** and **q.12** by adding phenylene-vinylene units (Table 2). The increase in cross section is

---

[13] Value obtained by rescaling the spectrum reported by Kennedy et al. [90] with the result at 585 nm, $\delta = 69$ GM, as reported by Fisher et al. [80]. This corrects for a typographical error in the earlier paper. See also Fig. 7.

[14] These values were obtained solving Eq. 4 for M_{ge} and using experimental results for all the other parameters of the equation, except Γ, which was assumed to be 0.1 eV for all compounds. This procedure is described in [94].

Table 2 2PA data for class I linear chromophores with phenylene-vinylene backbones (solvent: toluene)

Molecule	$\lambda_{max}^{(2)}$ (nm)	δ_{max} (GM)	Refs.
q.2	600	2.0×10^2	[94, 95]
q.10	730	1.0×10^3	[94, 95]
q.11	775	1.3×10^3	[94, 95]
q.12	810	1.5×10^3	[97]

rationalized by the fact that charge separation between the terminal donors and the center of the molecule leads to larger change in quadrupole moments in longer conjugated bridges, assuming that all other properties of the donor remain unchanged. For longer chain lengths, δ_{max} is expected to approach saturation, because the terminal groups could become decoupled. For the molecules in Tables 1 and 2, the parameter from Eq. (3) most directly related to the increase in cross section as a function of conjugation length is M_{ge}, while the value of M_{ef} is hardly affected [94]. The correlation between δ_{max} and M_{ge} is illustrated in Fig. 10 for diphenylpolyenes and D–π–D phenylene-vinylene oligomers with various conjugation lengths. It can be noticed that for each series of molecules the δ_{max} vs. M_{ge} data fall approximately on a straight line, except for the point corresponding to **q.9** (the longest diphenylpolyene), for which the increment in cross section with respect to the shorter molecules is significantly larger (and which probably reflects a change in the nature of the two-photon state [94]). It is also evident that the slope of the straight line in Fig. 10 is much larger in the phenylene-vinylene case than in the diphenylpolyene one, indicating that the degree of charge transfer for a given donor is mediated by the bridge.

Apart from its length, the influence of the conjugated bridge on $\lambda_{max}^{(2)}$ and δ_{max} is more complicated to describe, because, in general, M_{ge}, M_{ef}, E_{ge}, and E_{ef} depend on the nature of the bridge. A few examples of other conjugated skeletons or building blocks explored in the literature are (see Table 3): fluorene (**q.13**) and oligofluorenes [98], dithienothiophene (**q.14**) [99], dihydrophenanthrene (**q.15** and **q.16**) [100, 101], thiophene (**q.16**) [100], fused

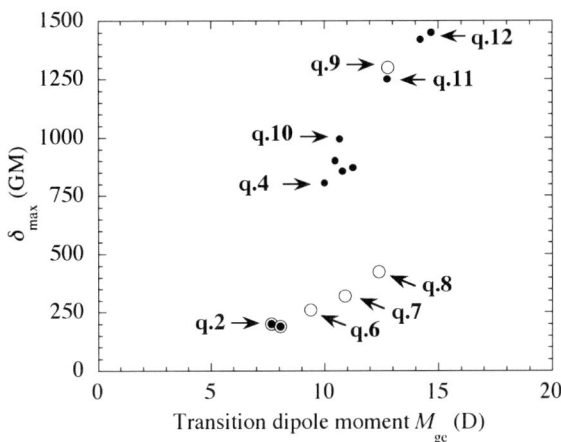

Fig. 10 Maximum 2PA cross section as a function of transition dipole moment M_{ge} for D–π–D diphenylpolyenes (*open circles*) and D–π–D distyrylbenzenes (*filled circles*). Data from [94, 97]. Data points for the molecules in Tables 1 and 2 are marked by the corresponding acronym

Table 3 2PA properties of quadrupolar molecules with various types of π-bridges. The solvent in which the measurement was performed is indicated in parenthesis after the molecular structure (#)

Molecule	$\lambda^{(2)}_{max}$ (nm)	δ_{max} (GM)	Refs.
q.13 (in chloroform)	534	55	[98]
q.14 (in 1,1,2,2-tetrachloroethane)	796*	2.7×10^2	[99]
q.15 (in dimethyl sulfoxide)	750	1.3×10^3	[101]

Table 3 (continued)

	Molecule	$\lambda_{max}^{(2)}$ (nm)	δ_{max} (GM)	Refs.
q.16	(in toluene)	740	2.6×10^3	[100]
q.17	(in toluene)	800	1.1×10^3	[102]
q.18	(in toluene)	780	5.4×10^2	[120]
q.19	(in toluene)	800	9.9×10^2	[120]

Table 3 (continued)

Molecule	$\lambda_{max}^{(2)}$ (nm)	δ_{max} (GM)	Refs.
q.20 (in dimethyl sulfoxide)	694	3.6×10^2	[103]

Some of the chromophores in this table are not strictly centrosymmetric, but their description as quadrupolar structures is a good approximation, based on the properties that have been observed or calculated.

* Single-wavelength measurement.

aromatic rings (q.17–19) [102], and ethynylene (q.18–20) [102, 103]. Among these, a particularly large cross section ($\delta_{max} = 2.6 \times 10^3$ GM [100]) was obtained for q.16, while chromophores containing triple bonds displayed modest cross sections [102, 103] when compared with phenylene-vinylene analogues with similar chain length. The type and length of the conjugated bridge can also influence the position of $\lambda_{max}^{(2)}$. In many cases, shifts in $\lambda_{max}^{(1)}$ lead to shifts in $\lambda_{max}^{(2)}$, but the magnitude of the shift can be hard to predict in centrosymmetric molecules, because the dependence of the excited state energy on molecular structure could be different for states e and f.

To summarize, the two general guidelines for increasing the 2PA cross section in quadrupolar molecules are: (1) increase the length of the molecule; (2) use electron donating and withdrawing substituents to facilitate intramolecular charge transfer. However, it should be noted that these changes also lead to changes in the position of the 2PA peak.

4.3
Dipolar Dyes

When electron-rich and -poor groups are substituted at the opposite termini of a π-backbone (class V in Fig. 8), an intramolecular charge transfer leads to the appearance of a dipole moment in the ground and/or excited states, hence the name *dipolar* for molecules of this type. Dipolar chromophores do not have an inversion center and one-photon allowed states can, therefore, be also two-photon allowed. This implies that the lowest-energy two-photon band can be observed for an excitation wavelength $\lambda_{max}^{(2)} = 2\lambda_{max}^{(1)}$, where $\lambda_{max}^{(1)}$ is the wavelength of the one-photon absorption band. Many of the studies devoted to the 2PA properties of dipolar chromophores focused on derivatives of E-4-amino-4'-nitrostilbene (d.1) and of N,N-diphenyl-7-[2-(4-pyridinyl)ethenyl]-9,9-di-n-decylfluoren-2-amine, a chromophore which has become known in the literature as AF-50 (d.2, Fig. 11).

The simplest description of compounds of this type is obtained limiting the sum-over-state equations to terms depending only on the properties of the ground state g and the first excited state e (two-state model) [93]. This is analogous to the two-level model introduced to describe other nonlinear optical properties, for example the nonlinear polarizability $\beta(-\omega; \omega_1,\omega_2)$ [104]. In the case of 2PA, this two-state, or dipolar, contribution to the cross section is, *on resonance*:

$$\delta = \frac{L^4}{5(\varepsilon_0 cn)^2 \hbar} \frac{M_{ge}^2 \left(\mu_e - \mu_g\right)^2}{\Gamma} \qquad \text{(SI units)} \qquad (5)$$

$$\delta = \frac{16\pi^2 L^4}{5c^2 n^2 \hbar} \frac{M_{ge}^2 \left(\mu_e - \mu_g\right)^2}{\Gamma} \qquad \text{(cgs units)}, \qquad (6)$$

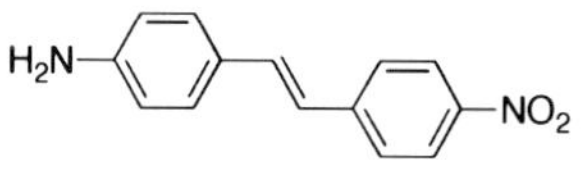

Fig. 11 Molecular structure of **d.1** and AF-50 (**d.2**)

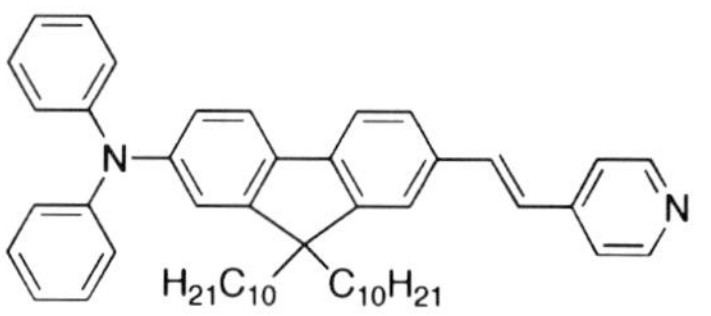

where μ_e and μ_g are the dipole moments of states e and g, respectively, and the other parameters have the same meaning as in Eqs. 3,4. The resonance condition in this case is given by: $E_{ge} = 2\hbar\omega$. For transitions to higher-lying excited states, the 2PA cross section contains, in general, a contribution of the type represented by Eqs. 5,6 and one of the type represented by Eqs. 3,4 (with state e being the intermediate state in the latter case).

In dipolar molecules, the 2PA cross section is expected to increase with conjugation length, because this leads to an increase in M_{ge} (as was the case for quadrupolar molecules) and to a decrease of the energy gap, E_{ge}. A few examples will be discussed below (Table 4). The cross section should also increase when the molecules exhibit a large *change* in dipole moment between the ground and excited state. The relationships between molecular properties, such as the degree of ground-state charge transfer and bond-length alternation (the difference between adjacent single C–C and double C=C bond lengths in polyene-type bridging groups) and values of dipole and transition dipole moments, have been discussed extensively in the past in the context of off-resonance values of the hyperpolarizabilities β and γ [105–108]. The design guidelines developed in those cases should also be applicable to 2PA, as the same fundamental quantities appear in Eqs. 5,6.

An early paper by Reinhardt et al. described the investigation of a large number of chromophores with general structure D-π-A for 2PA applications [109]. However, due to the fact that the measurements were carried out at a single wavelength and relied on the nonlinear transmission method using ns pulses, a method which is not selective for 2PA, it is not possible to extract detailed relationships between the structure of the chromophores and the position and magnitude of their 2PA band from that study. Nonetheless, some general trends for the *effective* cross section at 800 nm were identified with changes in the conjugation length of the molecule and the strength of the donor/acceptor substituents [109]. For compound **d.2** the cross section obtained from ultrafast measurements, which should approach the true 2PA cross section, is $\delta = 20$–30 GM at 790–800 nm [99, 110, 111].

The effect of conjugation length was addressed in various other studies, in which the cross sections were measured either by the 2P-induced fluorescence or by the z-scan methods with ps or fs pulses. For example (Table 4),

Table 4 2PA properties of dipolar molecules as a function of conjugation length. The solvent in which the measurement was performed is indicated in parenthesis after the molecular structure

	Molecule	$\lambda^{(2)}_{max}$ (nm)	δ_{max} (GM)	Refs.
d.3	H_2N—…—NO_2 (in dimethyl sulfoxide)	714	8.5	[112]
d.4	…—NO_2 (in dimethyl sulfoxide)	909	1.9×10^2	[113]
d.5	…N=N…—NO_2 (in dimethyl sulfoxide)	973	1.8×10^2	[113]
d.6	(F-substituted) (in dimethyl sulfoxide)	750	1.2×10^2	[115]
d.7	(F-substituted) (in toluene)	825	3.0×10^2	[115]
d.8	(F-substituted) (in toluene)	850	5.0×10^2	[115]

going from compound **d.3**, in which the bridge is a simple phenylene ring, to compound **d.4** with a stilbene bridge, δ_{max} increases from 8.5 GM [112] to 1.9×10^2 GM [113] (a similar cross section was obtained for the dibutylamino equivalent of **d.4** using the Kerr ellipsometry technique [114]). Approximately the same increase is obtained for an azobenzene bridge, as in compound **d.5**: $\delta_{max} = 1.8 \times 10^2$ GM [113].

In a series of diphenylpolyenes with a dimethylamino donor group attached to one phenylene ring and a penta-fluoro substitution on the other phenylene acting as an electron acceptor (**d.6–8**), the position of the 2PA peak shifts to longer wavelength and the cross section increases from 1.2 to 5.0×10^2 GM as the number of double bonds is increased from one to three (Table 4) [115]. This increase in cross section with the addition of double bonds is slightly larger for this series of dipolar diphenylpolyenes than for the centrosymmetric **q.2**, **q.6**, **q.7** discussed earlier (from 2.0 to 3.2×10^2 GM). However, compounds **d.6–8** have very low fluorescence quantum yields (2–3%), so their use in applications based on induced fluorescence could be limited.

The effect of substituent strength can be more difficult to predict in dipolar molecules, because the degree of charge separation in the ground and excited states can, in principle, affect the parameters M_{ge}, μ_e, and μ_g appearing in Eq. 5 in different ways and can influence the peak position through E_{ge}. Compounds **d.3**, **d.9**, and **d.10** are characterized by the same conjugated bridge (phenylene) and acceptor group (NO_2), but have different donors (see Table 5). The peak cross section is 8.5 GM in **d.3** [112], with an amine donor group, and about 5 GM in **d.9** [112], in which the donor, the maleimide functionality, is weaker than NH_2 [116]. However, the dipole moment μ_e is similar for the amino and maleimide derivatives, but the quantity ($\mu_e - \mu_g$) is actually larger in **d.9** [116]. Thus, in this case, the change in cross section from **d.3** to **d.9** appears to be due to other factors and the two-level description may not be adequate. Compound **d.10** also has a weaker donor than **d.3**, as the lone pair of the nitrogen can participate in the delocalization within the pyrrole ring. In this case, even if the lowest energy transition is significantly blue-shifted ($\lambda_{max}^{(1)} = 391$ nm in **d.3** and 291 nm in **d.10**), the 2PA cross section is actually larger in **d.10** ($\delta_{max} = 14$ GM) [112]. When the chain length is extended to azobenzene, there seems to be a more direct correlation between the substituent strength and the magnitude of δ_{max}. For example, replacing the dimethylamino group of **d.5** with the weaker maleimide as in **d.11**, the cross section decreases from ca. 1.8×10^2 GM [112, 113] to 84 GM [112] (and the peak position blue-shifts). A similar decrease is observed when the acceptor group of **d.5** (NO_2) is replaced by the weaker cyano group, with δ_{max} being 77 GM in **d.12** [113].

Table 5 2PA properties of dipolar molecules as a function of donor/acceptor strengths. The solvent in which the measurement was performed is indicated in parenthesis after the molecular structure. Compounds **d.3** and **d.5** from Table 4 are repeated here to facilitate comparisons

	Molecule	$\lambda_{max}^{(2)}$ (nm)	δ_{max} (GM)	Refs.
d.3	H_2N—⬡—NO_2 (in dimethyl sulfoxide)	714	8.5	[112]
d.9	(maleimide)—N—⬡—NO_2 (in dimethyl sulfoxide)	< 590	> 4.8	[112]
d.10	(pyrrole)N—⬡—NO_2 (in dimethyl sulfoxide)	700	14	[112]
d.5	N(CH₃)₂—⬡—N=N—⬡—NO_2 (in dimethyl sulfoxide)	973	1.8×10^2	[113]
d.11	(pyrrole)N—⬡—N=N—⬡—NO_2 (in dimethyl sulfoxide)	865	84	[112]
d.12	N(CH₃)₂—⬡—N=N—⬡—CN (in dimethyl sulfoxide)	944	77	[113]

4.4
Octupolar and Other Multibranched Dyes

Octupolar chromophores can offer an alternative design strategy for 2PA materials, and various structures belonging to this general category have been investigated. In many cases, these structures consist of a central unit on which electron rich or poor substituents and/or conjugated segments can be attached in a trigonal arrangement. Compounds of this type have been extensively studied in the past (and the interest continues to the present day) for nonlinear optical applications related to the second order susceptibility, $\chi^{(2)}$ [117]; since they lack an inversion center, these molecules can exhibit nonzero second-order polarizability $\beta(-\omega; \omega_1, \omega_2)$.

A prototype for octupolar molecules is crystal violet, **oc.1** (see Table 6). Beljonne et al. [118] have shown that **oc.1** has a 2PA peak for $\lambda_{max}^{(2)} = 752$ nm, with a cross section of 2.0×10^3 GM, a value in agreement with quantum chemical calculations presented in the same work. The magnitude of δ_{max} for this compound is quite remarkable, considering the relatively small size of the molecule. The band at 752 nm was assigned to the transition to the excited state 2A′, which is formally one-photon forbidden (but appears as a very weak feature in the absorption spectrum of **oc.1**), and two-photon allowed. The cross section for the transition to the lowest excited state (of 1E′ symmetry and one-photon allowed) was predicted to be about an order of magnitude smaller. Two-photon induced fluorescence was observed for **oc.1** in the wavelength range corresponding to this transition, $\lambda^{(2)} \approx 1200$ nm, but the cross section could not be quantified [118].

The most common choices for the core of octupolar chromophores for 2PA are 4,4′,4″-substituted triphenylamine and 1,3,5-substituted benzene. Cho et al. [119] have investigated the effect of chain length and donor groups (dialkylamine, diphenylamine, piperidine) in a series of compounds with a 2,4,6-substituted 1,3,5-tricyanobenzene core (**oc.2** in Table 6). It was found that, in general, δ_{max} increases with chain length. For example, in the case of R_1 branches, δ_{max} is in the range 2.0–3.0×10^2 GM, depending on the type of donor, while the range was 0.14–2.5×10^3 GM for the longer branches R_2 [119]. In all cases, this band corresponds to a transition into the lowest excited state of the molecule, $\lambda_{max}^{(2)} \approx 2\lambda_{max}^{(1)}$, which is both one-photon and two-photon allowed in noncentrosymmetric molecules such as these. However, for the molecule with the longest branch, R_3 (only one type of donor was considered in this study, D = N(phenyl)$_2$), δ_{max} (1.6×10^3 GM) in the same wavelength range is smaller than for the shorter analogue, R_2, with the same donor end group ($\delta_{max} = 2.5 \times 10^3$ GM).[15] The reduction in cross section was assigned to an increase in the conformational disorder of the structure going

---

[15] However, for molecule **oc.2** with R_3 and diphenylamine donors the largest cross section ($\delta_{max} = 2.6 \times 10^3$ GM) was observed at 800 nm, for the transition into a higher-lying state [119].

Table 6 2PA properties of octupolar and multibranched compounds. The corresponding linear building blocks are included for comparison, when available. The solvent in which the measurement was performed is indicated in parenthesis after the molecular structure

Molecule	$\lambda_{max}^{(2)}$ (nm)	δ_{max} (GM)	Refs.
oc.1	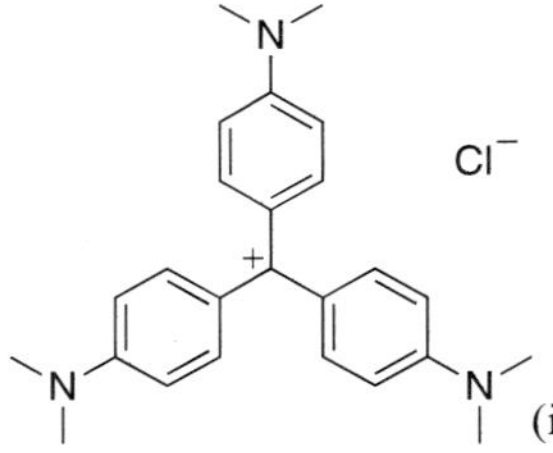 752	2.0×10^3	[118]
	(in glycerol)		
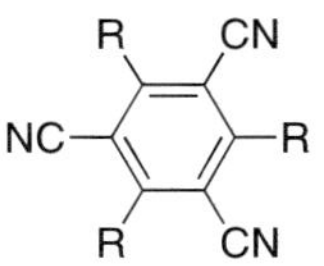 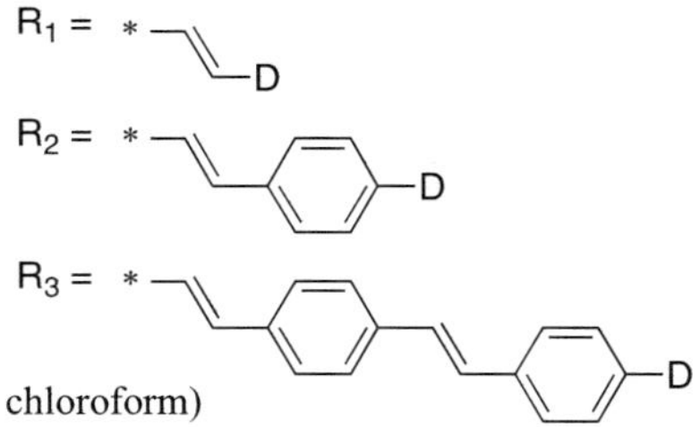 oc.2	See text	See text	[119]

Cl^-

$R_1 = *\!\!\diagdown\!\!D$

$R_2 = *\diagdown\!\!\diagup\!\!-\!\!D$

$R_3 = *\diagdown\!\!\diagup\!\!-\!\!D$

D = donor group (in chloroform)

Table 6 (continued)

Molecule		$\lambda^{(2)}_{max}$ (nm)	δ_{max} (GM)	Refs.
oc.3	(in toluene)	990 780	3.5×10^2 3.7×10^2	[120]
oc.4	(in toluene)	990 780	8.2×10^2 8.4×10^2	[120]
oc.5	(in toluene)	770	3.8×10^2	[121]
oc.6	(in toluene)	840	1.4×10^3	[122]

Table 6 (continued)

Molecule	$\lambda_{max}^{(2)}$ (nm)	δ_{max} (GM)	Refs.
oc.7	840	3.1×10^3	[122]
oc.8	840	5.1×10^3	[122]
oc.9	770	90	[123]

Table 6 (continued)

Molecule	$\lambda^{(2)}_{max}$ (nm)	δ_{max} (GM)	Refs.
oc.10 (in toluene)	815	2.0×10^2	[123]
	740	4.2×10^2	
oc.11 (in toluene)	815	2.9×10^2	[123]
	705*	$1.0 \times 10^{3*}$	

* Not a peak. This was the shortest wavelength investigated.

from R_2 to R_3 [119]. Compound **oc.4** has the same core as the **oc.2** series, but a different type of branch that includes ethynylene instead of vinylene units and is longer overall. Two peaks are observed in the 2PA spectra [120]. The one at 990 nm corresponds to the transition into the lowest excited state and has $\delta_{max} = 8.2 \times 10^2$ GM, the other is observed at 780 nm, in a region where the one-photon absorption spectrum does not have distinct features, and has comparable cross section. Comparing the results for **oc.4** and **oc.2**, it can be observed that conjugated linkers containing triple bonds are less effective than those containing double bonds in enhancing 2PA properties, as also discussed above for quadrupolar chromophores (Sect. 4.2). In order to assess the effect of the number of branches attached to a common core on the magnitude of the cross section, Yang et al. also studied compound **oc.3**, where only one branch and one acceptor group (CN) are present [120]. The 2PA spectrum has similar shape to that of **oc.4**, with $\delta_{max} = 3.5 \times 10^2$ GM at 990 nm. It was concluded that, for this molecular design, combining three linear chromophores of the type **oc.3** in the octupolar structure **oc.4** does not lead to an increase in δ_{max}, when this quantity is normalized to the number of branches (in fact, 8.2×10^2 GM/3 $= 2.7 \times 10^2$ GM, which is slightly smaller than δ_{max} of **oc.3**).

The degree of intermolecular charge transfer can be expected to be smaller in compound **oc.5** than in compounds **oc.2** and **oc.4**, because **oc.5** does not have the three cyano substituents. The 2PA spectrum of **oc.5** shows a broad band around 770 nm, with $\delta_{max} = 3.8 \times 10^2$ GM [121]. Even if a direct comparison with the **oc.2** series [119] is not possible because of the different nature of the conjugated linker and of the solvent used in the experiment, the cross section of **oc.5** is significantly smaller than for **oc.2** with the R_2 branch (which is shorter than the branch in **oc.5**) and a dialkylamino substituent ($\delta_{max} = 1.4 \times 10^3$ GM [119]).

Table 6 contains two other examples of multibranched structures consisting of one, two, or three branches around a common electron-rich triphenylamine core. In the first case, **oc.6–8**, the linear units belong to class III of Fig. 8 and share one of the terminal amine groups[16] [122]. The 2PA spectra of these three compounds, studied by Yoo et al., exhibit a peak at the same wavelength ($\lambda_{max}^{(2)} = 840$ nm), while a small red shift was observed in the position of the one-photon absorption band (473, 492, and 495 nm for **oc.6**, **oc.7**, and **oc.8**, respectively) [122]. The authors also reported an enhancement in the 2PA cross section going from one to three branches, by a factor larger than the increase in the number of branches (the ratio of δ_{max} for **oc.6**, **oc.7**, and **oc.8** are 1:2.3:3.7, while the number of branches increases as 1:2:3).

The second series of molecules, **oc.9–11**, was described by Katan et al. and is characterized by arms with a D–π–A structure (class V of Fig. 8), where the

---

[16] Compound **oc.6** is the same as **q.5**.

acceptor is a sulfonyl group and stilbene the linker [123]. It was found that in all three molecules the lowest energy 2PA peak is located at $\lambda_{max}^{(2)} \approx 2\lambda_{max}^{(1)}$ and has a band shape very close to that of the corresponding band in the 1PA spectrum, as expected for non-centrosymmetric molecules (but that **oc.10** and **oc.11** have a second, stronger peak at higher energy). The cross sections for this band are 0.90, 2.0, and 2.9×10^2 GM for **oc.9**, **oc.10**, and **oc.11**, respectively, and the increase with the number of branches is 1:2.2:3.2 [123]. These numbers indicate either a very small cooperative enhancement ($3.2/3 = 1.07$) in the cross section when the number of branches is increased or a simple additivity of the properties of the linear units (because the uncertainty in the measurement of δ for this study was reported to be 10%). However, the authors point out that metrics other than the mere number of branches can be used to compare the results and assess the cooperativity of the branches. For example, if the molecular weight of the molecules is used as normalizing factor, compounds **oc.10** and **oc.11** display an enhancement in δ_{max} of about 1.5 with respect to **oc.9** [123]. Alternatively, the cross sections can be normalized to the effective number of π-electrons in the system, N_{eff}, as defined by Kuzyk [124]. In this case, the value of δ_{max}/N_{eff} increases significantly going from **oc.9** to **oc.10** and to **oc.11** (5.0, 8.3, and 10 GM, respectively) and the enhancement is even larger when other spectral regions are compared [123].

4.5
Dendrimers

Dendritic structures could present potential advantages in various applications with respect to isolated molecules and polymers, because large chromophoric densities can be achieved by increasing the dendrimer generation, they are monodisperse macromolecules, and their design is flexible and could include multiple functionalities. In a few cases, dendrimers have been synthesized and investigated specifically for use in 2PA applications. We will discuss here the 2PA properties of two classes of dendrimers: (i) structures in which chromophoric units are linked together by branching groups that are not fully conjugated; and (ii) structures in which the branching points are integral part of the design of 2PA chromophores.

In the first case, dendrimers containing the chromophoric unit **dm.1** (see Fig. 12) on the periphery were synthesized up to generation 3 (in each generation the number of chromophore units doubles, so that generation 3 contains 8 chromophores) [125]. The 2PA cross section at 796 nm was found to increase by approximately a factor of 2 going from one generation to the next. It should be mentioned that the measurements were performed at a single wavelength in this case. However, due to the fact that the band in the 1PA spectrum does not exhibit a significant shift with generation number, the trend in δ values at 796 nm is likely to be similar to that of δ_{max}.

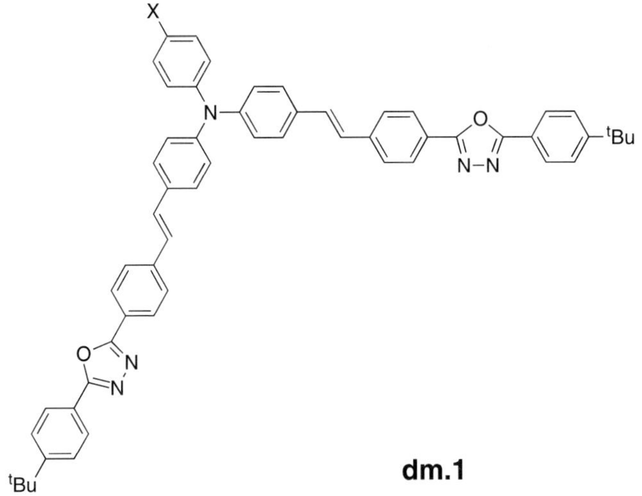

Fig. 12 Molecular structure of the chromophoric unit in the dendrimer investigated by Adronov et al. [125]. X is the attachment point to the dendrimer

In the second case, the branching point of the dendrimers is usually a tertiary amine, which also serves as a donor group for 2PA dyes belonging to class I of Fig. 8. Drobizhev et al. investigated dendrimers based on this scheme and having stilbene or distyrylbenzene moieties as the conjugated groups between branching points [126–128]. It was shown that δ_{max} initially scales with N^2, where N is the number of triphenylamine units in the dendrimer,[17] and the position of the band red-shifts, up to the compound with $N = 6$ (**dm.2–4**, see Table 7) [127]. This behavior was attributed to interchromophore coupling and the extension of the π-conjugation throughout the dendrimer, leading to a cooperative enhancement of the 2PA cross section [127]. For larger dendrimers, though, δ_{max} increases only linearly with N and $\lambda_{max}^{(2)}$ does not change further [126, 127]. For compound **dm.6**, with $N = 30$, δ_{max} at 690 nm was found to be 1.1×10^4 GM. The change in trend was attributed to an increase in conformational disorder in the larger dendrimers in the series, specifically to deviation from planarity between adjacent units.[18] When the linking unit between branching points in the dendrimer is changed from stilbene to distyrylbenzene, the peak cross section is found to increase at a lower rate for N between 2 ($\delta_{max} = 1.9 \times 10^3$ GM) and

---

[17] The triphenylamine group was chosen as the element on which to measure the size of the dendrimers in this series, because it was found that the 1PA oscillator strength is proportional to their number [127].

[18] It should be mentioned, however, that if the size of the dendrimer is gauged not by N, but by the number of diarylaminostilbene units ($N - 1$), the data can be fitted reasonably well by a straight line going through the origin, only the compound **dm.4** lying significantly more than one standard deviation from the line (if the 10% uncertainty reported in [126] can be extended also to the data described in [127]).

Table 7 2PA properties of dendrimers with triphenylamine branching points in dichloromethane solutions

Molecule		$\lambda_{max}^{(2)}$ (nm)	δ_{max} (GM)	Refs.
dm.2		670	3.2×10^2	[127, 128]
dm.3		680	1.3×10^3	[127, 128]

Table 7 (continued)

Molecule	$\lambda_{max}^{(2)}$ (nm)	δ_{max} (GM)	Refs.
dm.4	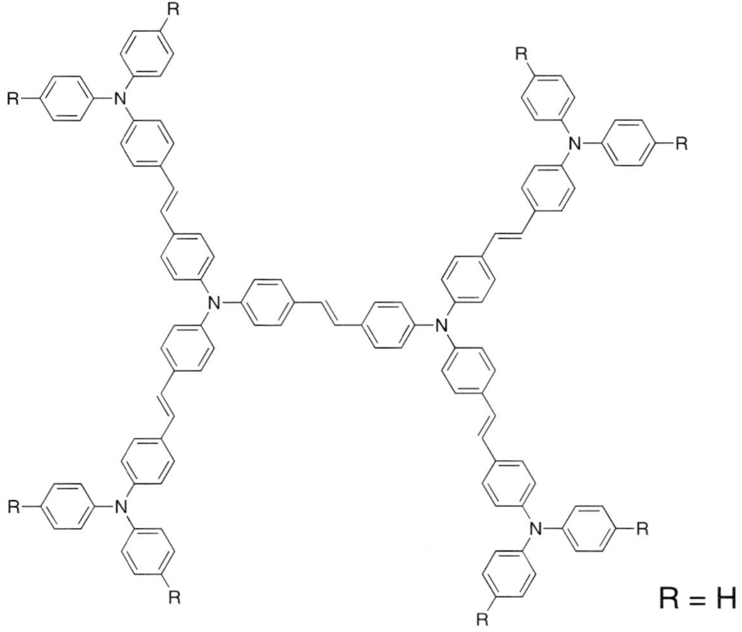 694	2.7×10^3	[127, 128]

R = H

Table 7 (continued)

Molecule		$\lambda^{(2)}_{max}$ (nm)	δ_{max} (GM)	Refs.
dm.5	" 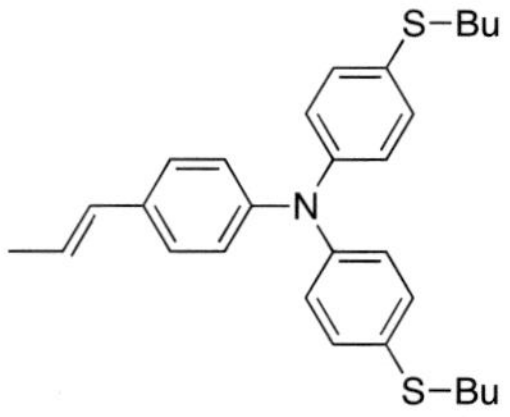	694	4.5×10^3	[127, 128]

R =

Table 7 (continued)

Molecule	$\lambda_{max}^{(2)}$ (nm)	δ_{max} (GM)	Refs.
dm.6 R =	690	1.1×10^4	[127, 128]

6 ($\delta_{max} = 9.9 \times 10^3$ GM), following a power law of the type $N_{\pi}^{1.35}$, where N_{π} is the total number of π-electrons in the dendrimer[19] [128].

In conclusion, for most of the molecules discussed here and for others reported in the literature (and for which full spectral characterization is available), the cross section of multi-branched chromophores is either found to scale linearly with the number of branches or exhibit a small enhancement when the molecular size is increased (or the evidence for the enhancement depends on the choice of normalization factors). The enhancement factor can be larger if cross sections at the same wavelength are compared, instead of the δ_{max} values, if there is a change in the band shape or position. These factors could be relevant for applications that have limitations on the operational excitation wavelength.

4.6
Polymers

We have seen above that the 2PA cross section scales with the conjugation length of the chromophores (even if the specific rate of increase depends on the type of bridge and substitution pattern). Conjugated polymers can then be of interest for 2PA application, as relatively long chain and/or conjugation lengths can be obtained. One difficulty in the characterization of the 2PA properties of polymers is due to the polydispersity of the materials obtained under many of the available polymerization methods. The measured 2PA cross sections, as well as other physical properties, are weighted averages over the composition of the polymer. However, the type of average could depend on the method used to measure the cross section (for example, the 2PIF signal under typical experimental conditions is proportional to the sample concentration, while the change in transmittance in nonlinear transmission and z-scan measurement is usually not proportional to the concentration, unless that change is very small).

A systematic study of 2PA properties of polymers as a function of degree of polymerization and polymerization conditions is not yet available. Table 8 contains a few examples of polymers whose 2PA spectra have been investigated. In some cases, design strategies identified in low-molecular weight molecules have been incorporated in the polymer structure. For example, pol.3 contains unit of the type D–A–D (class III in Fig. 8), where triphenylamines are the donor groups, shared by two adjacent units, and cyano groups are the acceptors. In this case, δ_{max} per repeat unit was found to be 1.0×10^3 GM [129], a value slightly smaller than for monomer q.12 with the same distance between donor groups, even without the acceptors (see

---

[19] However, counting again the number of chromophore units, instead of the π-electrons or the amine groups, the increase in δ_{max} is approximately linear (the ratio of cross sections between the largest dendrimer in the series, containing 5 distyrylbenzene units, and the parent compound, with one unit, is 9.9×10^3 GM$/1.9 \times 10^3$ GM = 5.2).

Table 8 2PA properties of selected polymers. The solvent used in the measurement is indicated in parenthesis after the molecular formula and the average number of repeating units. The cross sections reported are the contributions per repeat units

Molecule		$\lambda_{max}^{(2)}$ (nm)	δ_{max} per unit (GM)	Refs.
pol.1	Hex Hex, $x \approx 60$ (in chloroform)	625	3.3×10^2	[131]
pol.2	$R_1 = C_6H_{13}$ $R_2 = C_{10}H_{21}$ $x = 12$ (in toluene)	769	6.0×10^3	[132]

Table 8 (continued)

Molecule	$\lambda^{(2)}_{max}$ (nm)	δ_{max} per unit (GM)	Refs.
pol.3 $x \approx 12$ (in toluene)	$\geq 890^*$	$\geq 1.0 \times 10^3$	[129]

Table 8 (continued)

Molecule	$\lambda_{max}^{(2)}$ (nm)	δ_{max} per unit (GM)	Refs.
pol.4 $x \approx 4.3$ (in chloroform)	990	1.5×10^4	[130]
pol.5 $x \approx 91$ (in tetrahydrofuran)	710	$\approx 2 \times 10^{2**}$	[133]

* 890 nm was the longest excitation wavelength in the measurement.
** Estimated from the value of the 2PA coefficient reported in the original paper and obtained in a Kerr ellipsometry experiment with photon energies 1.96 eV and 1.55 eV for the pump and probe beams, respectively. 710 nm is the equivalent excitation wavelength for a degenerate experiment.

Table 1). This could suggest that there is a relatively high conformational disorder in the polymer, that the adjacent chromophores cannot be regarded as independent, or that the effective donor strength of the amine is different in the isolated chromophores and in the polymer, where it is shared by two chromophoric units.

In the case of **pol.4**, δ_{max} per repeat unit, 1.47×10^4 GM, is significantly larger than for the corresponding monomer, which has a octupolar structure similar to the **oc.2** series (with aminostyryl units in 1, 3, and 5 positions and cyano groups in 2, 4, and 6 positions of the benzene core) and $\delta_{max} \approx 5.4 \times 10^3$ GM [130]. Other polymers with the same octupolar monomer but connected by non-conjugated linkers have δ_{max} comparable to that of the monomer [130].

On a per chain basis, large 2PA cross sections have been reported even for polymers without donor and/or acceptor substitution [131–133]. For the polyfluorene **pol.1**, the contribution per repeat unit, 3.3×10^2 GM [131], is much larger and the peak position, $\lambda^{(2)}_{max} = 625$ nm, is red-shifted with respect to a shorter analogue that contains only two repeat units, 2,2'-(9,9-dihexyl)bifluorene, (δ_{max} per unit = 28 GM, $\lambda^{(2)}_{max} = 534$ nm) [134]. These results are consistent with a cross section that scales with x^2 for short chain lengths and with x for long chains, the transition between the two regimes occurring for $x \approx 12$ [131]. In polymer **pol.2**, which has a repeat unit with a length similar to that in **pol.1**, but more rigid, the contribution per unit is even larger, 6.0×10^3 GM [132]. The reduced conformation disorder in this ladder-type polymer is also reflected in the shape of the 2PA spectrum, which exhibits a clear vibronic structure, with two peaks separated by 0.16 eV [132].

4.7
Cyanines, Squaraines and Derivatives

Cyanines and squaraines are another class of linear conjugated molecules that have received attention for their peculiar nonlinear optical properties. Even though these molecular types are almost quadrupolar and quadrupolar, respectively, their properties are distinct from those of the quadrupolar chromophores discussed in Sect. 4.2. For example, cyanines exhibit small bond length alternation (at least for chain lengths below a limiting value) and have been predicted [135, 136] and found experimentally [137–139] to have off-resonance hyperpolarizability γ values that are negative and that scale as a function of chain length by a power law with an exponent larger than in other classes of conjugated molecules [140]. The ground state of squaraines, like that of cyanines, can also be represented in terms of two degenerate valence-bond structures (for squaraines, these are characterized by a large degree of intramolecular charge transfer between the central electron poor squarylium group and the electron rich end groups), and can show negative off-resonance second-order hyperpolarizability [136]. Information on the position of two-photon allowed states for some chromophores belonging to these classes have been deduced from studies of other components of γ, for example the one relevant in the third-harmonic generation process, as the presence of two-photon resonances can modify the frequency dependence and the sign of γ [141, 142].

As discussed above, the degree of intramolecular charge transfer is related to the magnitude of the 2PA cross section, and, as such, cyanines and squaraines could also have interesting 2PA properties.

In compounds **cy.1, cy.2, sq.1, sq.2** (Table 9) the lowest energy two-photon transition is observed just above (≈ 0.15 eV) the one-photon state, close in energy to the shoulder observed in the 1PA spectra of these compounds [143, 144]. The nature of this two-photon state has not yet been unambiguously assigned. According to Scherer et al., the final state of this transition is a vibronic level of the lowest B_u state of the molecule, with the transition being allowed through vibronic coupling only [144]. This is supported by the observation that theoretical calculations on other cyanines predict the lowest two-photon state to be at a much higher energy (> 0.5 eV) [135]. However, the relatively large cross section for this 2PA transition (2.0×10^2 GM for **sq.1** [144] and 1.4×10^2 GM for **cy.2** [145] at 1180 nm, see Table 9) and significant transition dipole moment M_{ef} estimated from other nonlinear optical measurements (e.g., 4.8 D for the squaraine usually referred to as ISQ [141], which has an indoline derivative as donor group) could suggest a fully allowed 2PA state. The next 2PA band is observed 1.1 eV and 0.57 eV above the one-photon state for **sq.1** and **sq.2**, respectively, and is also present in longer oligomers [144]. The 2PA cross section is larger for this transition than for the lower energy one and increases with the length of the conjugated chain ($\delta_{max} = 3.3 \times 10^2$ GM for **sq.1** and 4.7×10^3 GM for **sq.2**), mainly because of the increase in the transition dipole moment M_{ge}, but the contribution per repeating unit levels off after the trimer [144]. Similar spectral features are observed in **cy.2**, for which the cross section for the higher excited state is 7.2×10^2 GM, five times that for the transition at 1180 nm [145].

In the case of the pseudocyanine **cy.3**, however, the lowest energy 2PA transition was observed for $\lambda_{max}^{(2)} = 760$ nm (no 2PA was observed near $2\lambda_{max}^{(1)}$), with $\delta_{max} \approx 5.0 \times 10^2$ GM [146]. It was also observed that when this compounds forms aggregates, a large increase in the 2PA cross section takes place below 640 nm, δ_{max} increasing from 1.0×10^3 to 1.0×10^4 GM at 600 nm (because the detuning term is smaller and the transition moment M_{ge} larger for the aggregate than the isolated molecules), while only a small change occurs near the 760 nm peak [146].

Very large cross sections (up to 3.3×10^4 GM) have been reported by Chung et al. for a series of squaraines that differ from **sq.1–2** in the type of terminal donor and conjugated linker [147]. Two examples are compounds **sq.3** and **sq.4** in Table 9. The state corresponding to this transition lies about

Table 9 2PA properties of squaraines and cyanines. The solvent in which the measurement was performed is indicated in parenthesis after the molecular structure

Molecule	$\lambda_{max}^{(2)}$ (nm)	δ_{max} (GM)	Refs.
cy.1 (in dimethylformamide)	≈ 1030	–	[143]
cy.2 a) R = CH$_3$ (in dimethylformamide)	(a) ≈ 1180	(a) –	[143]
b) R = C$_3$H$_7$ (in ethanol)	(b) 1180 780	(b) 1.4×10^2 7.2×10^2	[145]
cy.3 (in 1:1 water/dimethyl sulfoxide mixture)	760 620*	$\approx 5 \times 10^2$ $\approx 1.0 \times 10^3$	[146]
aggregate of **cy.3** (in 1:1 water/dimethyl sulfoxide mixture containing NaCl)	760 620*	8×10^{2}** 1.0×10^4 **	[146]

Table 9 (continued)

Molecule	$\lambda^{(2)}_{max}$ (nm)	δ_{max} (GM)	Refs.
sq.1 a) R = C_2H_5 (in chloroform) b) R = C_3H_7 (in ethanol)	(a) 1181 821 (b) 1150 820	(a)# 2.0×10^2 3.3×10^2 (b) 40 $\approx 7 \times 10^2$	[144] [145]
sq.2 (in chloroform)	1038 ≈ 1240	4.7×10^3 $\approx 1.4 \times 10^3$	[143, 144]
sq.3 (in dichloromethane)	960	2.7×10^4	[147]

Table 9 (continued)

Molecule	$\lambda_{max}^{(2)}$ (nm)	δ_{max} (GM)	Refs.
sq-4 (in tetrahydrofuran)	1050	3.3×10^4	[147]

* Shortest excitation wavelength in the study.

** Referred to the concentration of monomer units.

Values for Table 2 in the reference cited (the relative intensity between the two peaks is different in Fig. 4 from the same source).

0.9 eV above state e and has been identified as state $3A_g$, using the results of quantum chemical calculations (in this study, the energy range extending down to state e had been specifically investigated only for compound **sq.3**, and no significant 2PA activity was detected) [147]. Using a three-state equation equivalent to Eq. 3, the authors ascribe the large cross sections of these chromophores to the large values of M_{ge} and the small detuning energies, $(\Delta E = 0.2-0.3 \text{ eV})$ [147]. When compared to substituted distyrylbenzenes with comparable conjugation length (see, for example, **q.10–12** in Table 2), for which $\Delta E \approx 1.2$ eV, the detuning term alone can be responsible for an enhancement in δ_{max} for **sq.3–4** of an order of magnitude. However, the comparison with the quadrupolar chromophores discussed earlier or other quadrupolar dyes is not straightforward from a structure/property point of view, because in the distyrylbenzene case the transition investigated was typically only that to the $2A_g$. In addition to a possible difference in the nature of the $2A_g$ and $3A_g$ states, distyrylbenzenes could also have higher lying 2PA-allowed states that may "benefit" from a small detuning energy. For example, there is evidence that a high-lying totally symmetric state, mA_g, can be strongly coupled with state e [135, 148, 149]. A large transition moment for this state in conjunction with a small ΔE would lead to very large cross sections also in quadrupolar molecules. On the other hand, the relatively sharp onsets of 1PA in squaraines can allow one to probe more readily states with small detuning energies.

4.8
Macrocycles (Porphyrins and Phthalocyanines)

Porphyrins, phthalocyanines, and related larger macrocycles are characterized by a framework of conjugated π-electrons which extends in two dimensions and they represent an alternative type of building block for the study of nonlinear optical properties. In addition to interest arising from the involvement of chromophores in this category in photosynthesis and other natural processes, tetrapyrrolic compounds are currently of interest in photodynamic therapy (where they are used as photosensitizers for singlet oxygen generation, see Sect. 5.6) and in optical pulse suppression (because of the large cross section for triplet-triplet transitions in spectral regions of weak linear absorption) [150]. Their 2PA properties have only recently been addressed. Here we will primarily discuss reports on porphyrins (see Table 10), although other macrocycles have also been investigated.

It has been shown that the 2PA cross section of porphyrin derivatives is relatively small, given the large size of the macrocycle, in the spectral range $\lambda_{max}^{(2)} \approx 2\lambda_Q^{(1)}$, where $\lambda_Q^{(1)}$ is the wavelength at which the Q band is observed by 1PA spectroscopy [151]. For example, compound **mac.1** exhibits a weak band with $\delta_{max} = 4$ GM at 1150 nm, corresponding to a state energy about 1300 cm^{-1} higher than the lowest energy Q band [152]. For other por-

Table 10 2PA properties of tetrapyrrolic derivatives. The solvent in which the measurement was performed is indicated in parenthesis after the molecular structure

Molecule	$\lambda_{max}^{(2)}$ (nm)	δ_{max} (GM)	Refs.
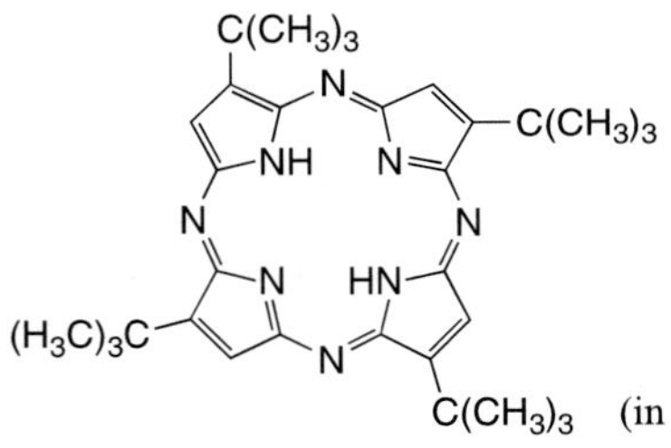 mac.1 (in dichloromethane)	1150 715*	4 85*	[151, 152]
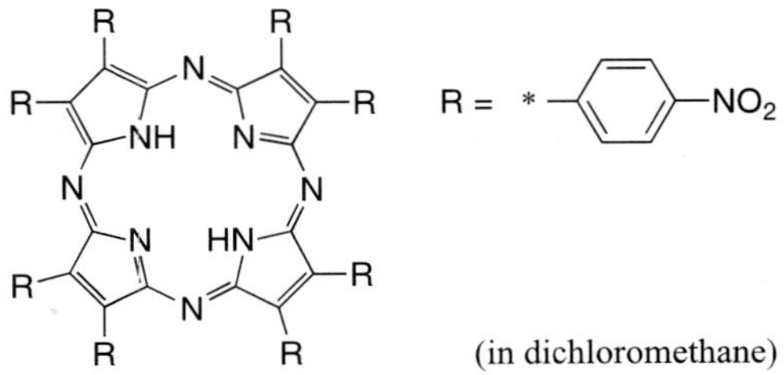 mac.2 (in dichloromethane)	1220 770*	4 1.6 × 10³*	[151, 152]

C(CH₃)₃

C(CH₃)₃

(H₃C)₃C C(CH₃)₃ (in dichloromethane)

R = * NO₂

R (in dichloromethane)

Table 10 (continued)

Molecule	$\lambda_{max}^{(2)}$ (nm)	δ_{max} (GM)	Refs.
mac.3 (in dichloromethane + 1% pyridine)	830	20	[154, 155]
mac.4 (in dichloromethane + 1% pyridine)	830	8.5×10^3	[154, 155]

Table 10 (continued)

Molecule	$\lambda_{max}^{(2)}$ (nm)	δ_{max} (GM)	Refs.
mac.5 R = C$_7$H$_{15}$ (in chloroform)	873	1.8×10^3	[156]
mac.6 R = C$_7$H$_{15}$ (in chloroform)	887	7.6×10^3	[156]

Table 10 (continued)

	Molecule	$\lambda^{(2)}_{max}$ (nm)	δ_{max} (GM)	Refs.
mac.7		1325	1.15×10^5	[158]

R' = H   (C=O)N(CH$_2$CHEt(Bu)$_2$)

R = *   (C=O)N(CH$_2$CHEt(Bu)$_2$)

N = 13 ± 3 (in carbon tetrachloride) (#)

* Not a peak; this was the shortest $\lambda^{(2)}_{max}$ investigated. See text for wavelength dependence of δ in this region.

See [158] for definitions of R and R' for other values of N.

phyrins, for example, 5,10,15,20-tetraphenylporphine, a weak 2PA activity ($\sim$1 GM) can also be seen at the energy of the Q band, possibly indicating a lowering of the symmetry of the chromophores to a non-centrosymmetric one [151].

Drobizhev et al. have also measured the 2PA spectra in the 700–800 nm range and shown that the 2PA cross section is larger by approximately an order of magnitude here than near $2\lambda_Q^{(1)}$. For example, δ = 85 GM at 715 nm, for **mac.1** [152]. In many of the chromophores originally investigated, this range includes or approaches $\lambda^{(2)} \approx 2\lambda_B^{(1)}$, where $\lambda_B^{(1)}$ is the one-photon excitation wavelength for the B band. In general, the 2PA spectrum has not been found to follow the shape of the corresponding 1PA band or to exhibit a distinct peak. Instead, the cross section increased with decreasing wavelength and the behavior was explained as an effect of the decrease in the detuning energy ΔE (see Eq. 4) when $\hbar\omega$ approaches a one-photon resonance, which, in the case of porphyrins, is that associated with the Q band [151, 152]. The authors have shown that, using three-level equations similar to Eq. 4, it is possible to factor out the terms affected by the resonance enhancement and assess the existence of a transition to an A_g state in this energy range [151]. Cross sections of the same order of magnitude as **mac.1** have been recorded for alkyl- and alkoxy-substituted phthalocyanines in the Q and B band regions [153].

When the periphery of the tetrapyrrolic cycle bears electron withdrawing groups, as in **mac.2**, instead of alkyl substituents, as in **mac.1**, the cross section in the B band region increases further (δ = 1.6 $\times$ 10^3 at 770 nm for **mac.2**), because of a decrease in ΔE and an increase in M_{ef}, and the cross section was found to be proportional to the Hammett constant of the substituents [152].

Another strategy pursued for increasing the cross section in tetrapyrrolic materials has been that of extending the length of the chromophores by linking macrocycles by conjugated bridges. For example, in the B band region, the cross section increases by over two orders of magnitude going from the monomer **mac.3** (δ_{max} = 20 GM) to a dimer with an acetylene linker, **mac.4** (δ_{max} = 8.5 $\times$ 10^3 GM) [154, 155]. In this case, a clear peak can be seen for all the molecules in the series at a transition energy on the blue side of the B band. In the same study it was shown that the magnitude of the increase depends on the nature of the conjugated bridge, and that longer linkers do not always lead to larger cross sections. In this series of molecules, the three-level model of Eq. 4 was again used to help in the interpretation of the results. It was shown that, while all the parameters in Eq. 4 contribute to the enhancement in the cross section going from **mac.3** to **mac.4**, the largest contribution actually comes from M_{ef} [154]. This is in contrast with results on quadrupolar chromophores (see Fig. 10), for which a change in chain length mostly affects the value of M_{ge}.

The dimer units of **mac.6** have been designed [156] such that the 1-methyl-imidazolyl group in the *meso* position of one molecule can coordinate the metal center in another unit, forming a supramolecular dimer (the second porphyrin unit in the building block is a free base). If 1-methylimidazole, a competitive ligand, is added to the solution in sufficient concentration, the dimers can be dissociated into the monomer **mac.5** (which itself contains two phorphyrin cycles covalently linked). It was found that the formation of the dimer has a positive effect on the 2PA cross section for a transition energy in the range of the B band, δ_{max} per porphyrin unit being 0.9×10^3 GM for **mac.5** and 1.9×10^3 GM for **mac.6** (the overall cross sections per molecular or supramolecular unit are included in Table 10) [156]. When a bis-zinc analogue of **mac.5** is used as a building block, the two porphyrins in each dimer can be coordinated by the terminal imidazolyl groups belonging to two distinct dimers, forming a polymer. The value of δ_{max} per porphyrin, 2.0×10^3 GM, is, however, comparable to that of **mac.6**, indicating that **mac.6** is already close to the saturation regime for this type of structure [156].

In a different supramolecular approach, zinc porphyrin oligomers with butadiyne linkers can form the ladder structure of **mac.7** when a bidentate ligand, such as 4,4'-bipyridyl, is used [157]. Oligomers with different chain lengths were synthesized and it was found that δ_{max}/N, where N is the number of macrocycles in the oligomers or in the ladder structure, decreases slightly going from the linear to the ladder structure up to $N = 4$, after which an increase is observed [158]. In all cases except $N = 2$, the formation of the ladder structure leads to a large red-shift in $\lambda_{max}^{(2)}$. For the longest oligomer in the series, δ_{max}/N was measured to be 8.9×10^3 GM, so that the total cross section for the structure is above 1×10^5 GM [158].

4.9
Organometallics

In organometallic complexes, the 1PA spectrum in the visible and near-infrared range is often dominated by a strong band assigned to charge transfer between the metal centers and the conjugated ligands. While a number of studies investigated the nonlinear properties of this type of materials at individual wavelengths [159–161], which may be of interest for specific applications, spectral information is available only on few compounds.

In the case of the platinum ethynyl compound **om.1**, for which $\lambda_{max}^{(1)} = 360$ nm, a broad 2PA band was observed around 630 nm, with $\delta_{max} \approx 3.0 \times 10^2$ GM [162, 163]. In dendritic-type compounds containing multiple metal centers but still with phenylene ethynylene ligands, larger 2PA cross sections were reported. For example, **om.3** displays $\delta_{max} \approx 3.5 \times 10^3$ GM for the peak at 740 nm, but no 2PA activity was detected into the lowest one-photon state, e ($\lambda_{max}^{(1)} = 469$ nm) [164]. In **om.2**, which has shorter inner branching unit and a lower number of metal centers, the magnitude of the cross section is approximately the same, but the peak is red-shifted with respect to **om.3**, and this corresponds to the transition into state e [165].

Compounds **om.4** and **om.5** have the same metal center as **om.1**, but longer conjugated linkers, which also incorporate fluorene units, and they have terminal donor groups. 2PA spectra of these molecules are characterized by two bands [166]. One was observed at $\lambda_{max}^{(2)} \approx 2\lambda_{max}^{(1)}$ (780–790 nm), with $\delta_{max} = 1.6$ and 3.0×10^2 GM for **om.4** and **om.5**, respectively, the other for $\lambda_{max}^{(2)} \approx 630$–640 nm, with a cross section about a factor of three larger [166]. For both bands the cross section is larger in the longer chromophore, **om.5**. The higher-energy band was assigned to the transition to a totally symmetric electronic state, as no strong features were observed in the 1PA spectrum in the 300–350 nm range. The appearance of the lowest energy band was instead interpreted as due to the presence of noncentrosymmetric conformers in solution [166].

5
Two-Photon Chemistry

5.1
Introduction

As discussed in the introduction, two-photon absorption offers the ability to create excited states with high spatial resolution in three dimensions. For some applications, the purely photophysical properties of these excited states, such as fluorescence or excited-state absorption, are important. Alternatively, the excited states populated by 2PA may be used to activate chemical processes with three-dimensional resolution. This can occur either through energy or electron transfer from the chromophore to another species, or by a direct chemical reaction involving the excited chromophore. In general, most organic chromophores obey Kasha's rule; i.e., regardless of which excited states are accessed by 1PA and 2PA, relaxation from these states typically occurs on a sub-ps timescale, leading to the population of the lowest vibrational level of the first singlet excited state (state e of Fig. 1, often denoted as S_1). Subsequent photophysics and photochemistry generally occur from S_1 (or from the lowest triplet state, T_1, reached from S_1 by intersystem crossing), and is the same independent of whether 1PA or 2PA is the initial excitation mechanism used. Thus, to some extent, one can design materials for 2P-induced chemistry based on systems with well-established 1P photochemistry while, in addition, incorporating structural features that can provide large 2PA cross sections.

Table 11 2PA properties of organometallic compounds. The solvent in which the measurement was performed is indicated in parenthesis after the molecular structure

Molecule	$\lambda^{(2)}_{max}$ (nm)	δ_{max} (GM)	Refs.
om.1 (in 1,2-dichloroethane)	630[#]	$\approx 3 \times 10^{2\#}$	[162, 163]
om.2 (in dichloromethane)	1000	$3.9 \times 10^{3*}$	[165]

Table 11 (continued)

Molecule	$\lambda_{max}^{(2)}$ (nm)	δ_{max} (GM)	Refs.
om.3	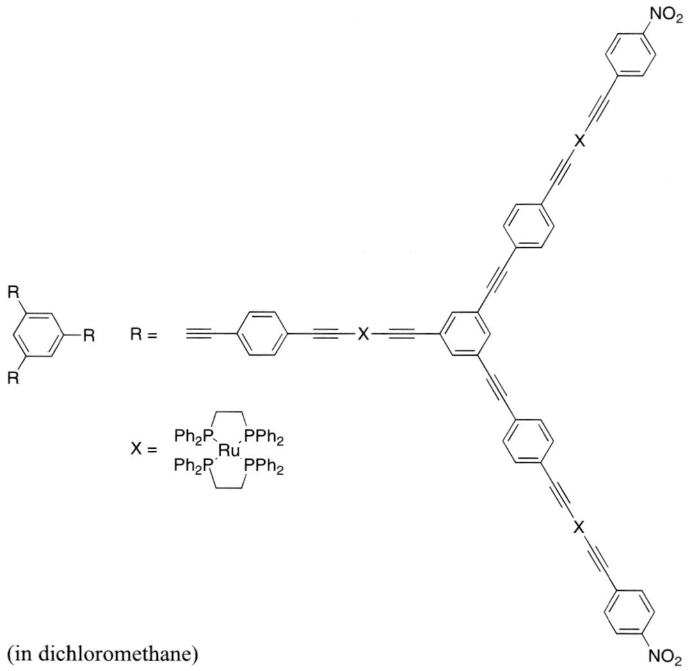 740	3.5×10^3	[164]

(in dichloromethane)

Table 11 (continued)

Molecule	$\lambda^{(2)}_{max}$ (nm)	δ_{max} (GM)	Refs.
om.4 (in benzene)	780 634	1.6×10^2 5.8×10^2	[166]
om.5 (in benzene)	790 644	3.0×10^2 9.6×10^2	[166]

Value at 630 nm determined from the spectral shape reported in [162] and the 2PA absorption coefficient at 595 nm reported in [163].

* Calculated from the value of Im γ reported in the cited paper, using a conversion equation consistent with other papers by the same authors.

As introduced in Sects. 1 and 2, 2P-induced chemistry has attracted interest for applications in two main areas: in medicine, where it can enable the generation or release of biologically active species in the desired region, even deep in tissues; and in the fabrication of three-dimensional microstructures from polymeric, metallic, biological, and hybrid materials. In this section, we will mainly focus on the chemistries initiated by two-photon and, more generally, multi-photon absorption that have been used in microfabrication: first we will consider the use of multi-photon absorption to initiate radical reactions and the use of these reactions in microfabrication (Sect. 5.2), followed by the use of multi-photon absorption to pattern polymers through photoacid generation (Sect. 5.3). We will also survey other material classes that have been patterned using multi-photon absorption (Sect. 5.4). For a more detailed account of the properties and applications of microfabricated structures, we refer the reader to other recent reviews [7, 9]. Finally, we will briefly consider the use of 2PA in the photorelease of a variety of functional groups (Sect. 5.5) and in generating singlet O_2 for biological and medical applications (Sect. 5.6).

5.2
Two-Photon-Induced Radical Chemistry

2P-induced radical chemistry may be achieved either through the direct creation of reactive intermediates from the photoexcited species, or following electron or energy transfer from the 2P-excited species to another moiety, which can then initiate radical reactions; both methods are being increasingly used for the microfabrication and nanofabrication of 3D objects, and have also been exploited as a means of creating high-density memory systems.

Resins containing acrylate monomers and/or oligomers are among those that can undergo 2P-induced radical polymerization. 2P-initiated polymerization of acrylates was first demonstrated using commercial negative-tone resists, i.e., materials which become less soluble on irradiation, incorporating "conventional" photoinitiators, i.e., initiators developed and used for 1PA-based applications [3, 4]. This work using conventional photoinitiators has been extended to the fabrication of complex three-dimensional microstructures by two-photon-induced polymerization.

A 1997 study by Kawata and co-workers used a commercial resin, SCR500 from Japan Synthetic Rubber Company, consisting of urethane-acrylate monomers, some of which contain more than one acrylate group, and two conventional photoinitiators, c.1 and c.2 (Fig. 13, top), and a 790 nm/200 fs laser [20]. These initiators are known to undergo photocleavage of the PhCO–R bond to give PhCO· and R· radicals upon irradiation; these radicals can then initiate acrylate polymerization. Due to the presence in the resin of species containing more than one polymerizable acrylate group, the polymerization

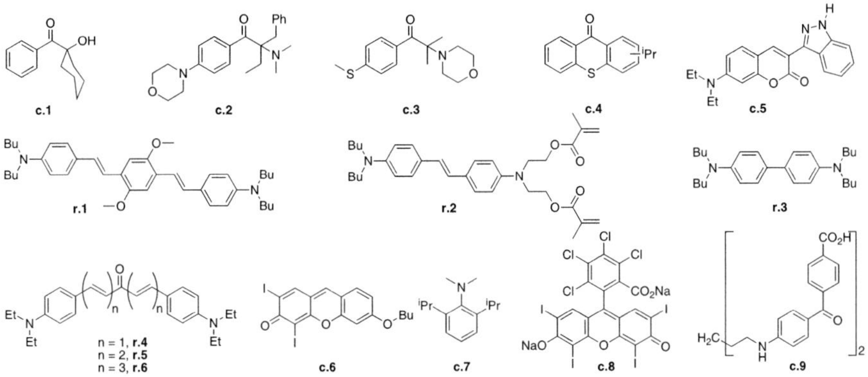

Fig. 13 Structures of compounds discussed in the context of 2P-initiated radical polymerization

process leads to crosslinking of the polymer chains and yields an insoluble material in the irradiated volume around the focus of the laser beam (this volume is often referred to as voxel). The film can then be developed using a solvent (ethanol, in this particular case) to remove the resin from the unexposed areas. The microstructure demonstrated in this study was a helical coil of 7 μm diameter and 50 μm length with a line cross section 1.3 μm × 2.2 μm (Fig. 14, top). Subsequent work by the same group [167, 168] using SCR500 includes fabrication of microscale model of a bull with a height of ∼7 μm and a length of ∼10 μm (Fig. 14, bottom) using a 780 nm/150 fs laser.

Many of these conventional initiators have rather low 2PA cross sections; indeed, a recent study of a range of conventional initiators, including **c.1**–**c.4** (Fig. 13, first row), reports that these chromophores have 2PA maxima at photon wavelengths ranging from 530 to 760 nm, and peak cross sections of between 5 and 23 GM, as obtained with the z-scan measurement technique [169]. This is consistent with the fact that initiators such as **c.1**–**c.4** have relatively small π-electron systems, which typically posses only modest 2PA cross section, as seen in Sect. 3. Microfabrication of a log-stack structure from acrylates with an 800 nm/120 fs laser has also been accomplished using a coumarin derivative, **c.5** (Fig. 13), the cross section of which has been determined using 2P-induced fluorescence to be ca. 30 GM at this wavelength, in conjunction with $[Ph_2I]^+[PF_6]^-$, as a 2P-initiator system [170]. The one-photon studies of this initiation system [171] suggest that photoexcited **c.5** transfers an electron to the diphenyliodonium ion to give $[Ph_2I]$, which then decomposes to give PhI and phenyl radical [170].

As discussed in Sect. 4, chromophores with significantly higher cross sections than **c1.**–**c.5** have been identified and developed in recent years. The use of photoinitiators with larger 2PA cross sections than the conventional ones can potentially permit the use of lower concentrations of initiator, faster fabrication speeds, and/or lower laser intensities. It was demonstrated that

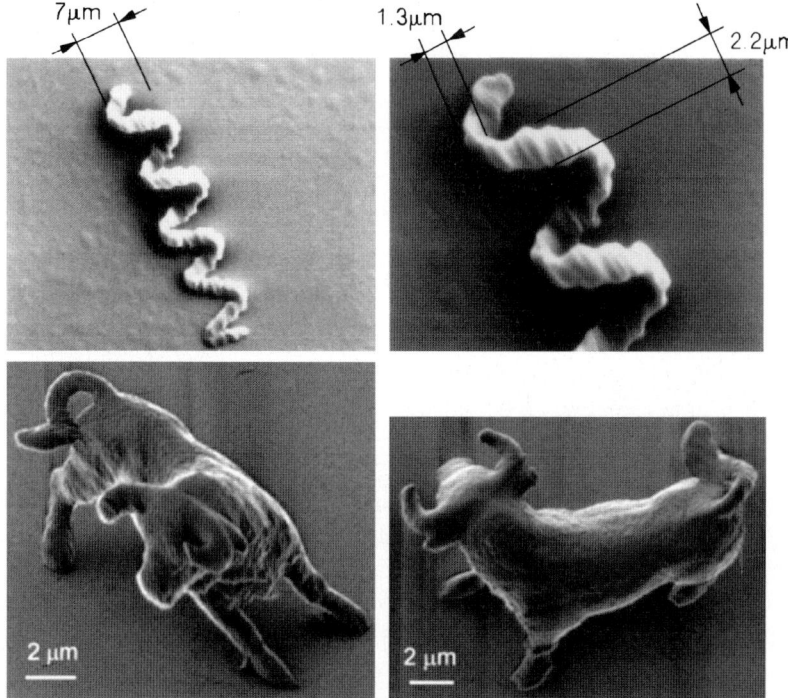

Fig. 14 Scanning electron microscopic (SEM) images of a spiral coil (*top*) [20] and a microscale bull (*bottom*) [7, 167] made by two-photon radical polymerization of the SCR500 resin. Reproduced from [7, 20] with kind permission of Springer Science and Business Media, and of the Optical Society of America

chromophores such as **q.2**, **q.11** (Table 2), and **r.1** (Fig. 13), which have peak cross sections of 200–1300 GM in the range 600–730 nm, can be used to initiate the polymerization of acrylate monomer via a radical process [21, 172]. These chromophores have been used in conjunction with crosslinkable triacrylates (SR9008 and SR368, from Sartomer) and an inert polymer binder for 2P-fabrication of a range of structures, including photonic crystals, waveguides, cantilevers, and interconnected chain links (see Fig. 15 for examples) [21, 172, 173]. Several subsequent studies of multiphoton radical polymerizations have also made use of specifically designed 2P chromophores [174–176].

The efficiency of the 2P-induced polymerization process in SR9008 using **q.2** as an initiator has been compared to that using conventional radical photoinitiators, including benzil, benzophenone, bis(*N*,*N*-dimethylamino)benzil, 4,4'-bis(*N*,*N*-dimethylamino)benzophenone, **c3**, and **c4**. The threshold power[20] for

---

[20] In this context, the threshold power is the minimum power necessary to obtain a well formed structure which is not deformed or washed away during the developing step. As the 2P-excitation rate does

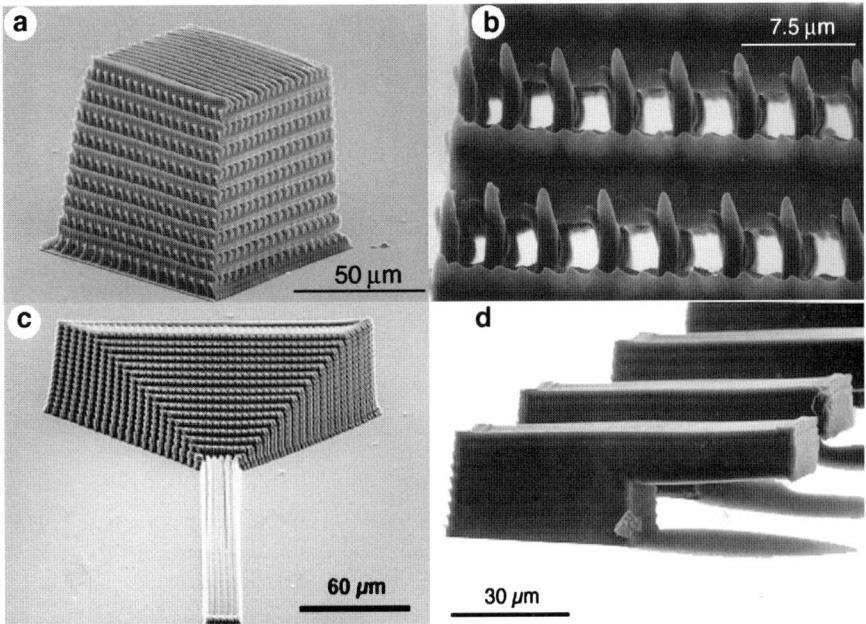

Fig. 15 Three-dimensional microstructures produced by 2P-initiated radical polymerization of triacrylates using **q.11** or **r.1** as initiators: **a** photonic bandgap structure; **b** close-up view of the structure in (**a**); **c** tapered waveguide structure; and **d** array of cantilevers. Reproduced with permission from [21]

polymerization at 600 nm with 5 nm pulses focused to a 2.5 μm spot size using **q.2** was found to be 30% of that using the most effective conventional initiators, **c.3**, and 4,4′-bis(N,N-dimethylamino)benzophenone [21, 172]. At the longer photon wavelength of 730 nm and using 150 fs pulses focused to a 0.35 μm spot size, **r.1** was found to show an order-of-magnitude improvement in the dynamic power range for polymerization of a resins based on monomers SR9008 and SR368 (that is the range between the threshold power and the power at which the material is damaged) relative to benzil, 4,4′-bis(N,N-dimethylamino)benzophenone, and **c.3**. This increased sensitivity depends on both the 2P cross section at the wavelength in question and on the efficiency of the radical generation process. The threshold power for compound **q.11** at 800 nm is about 50% larger than that for **r.1** at 730 nm [21, 177].

The mechanism for radical initiation by chromophores such as **q.2** is believed to begin with electron transfer from the excited chromophores to an acrylate group [21]. Electrochemical measurements suggest that this electron

---

not depend simply on the average or peak power, but more specifically on the local intensity of the beam (see Eq. 2), and, thus, on the beam focusing conditions, threshold values are only valid in relative terms and need to be obtained under the same excitation conditions for different resins.

transfer is energetically feasible; further support for intermolecular electron transfer from photoexcited **q.2** to acrylates comes from experiments indicating steady-state fluorescence quenching, fluorescence-lifetime shortening, and the appearance of absorption bands characteristic of **q.2**$^{\bullet+}$ [21, 172, 173]. Intramolecular photoinduced electron transfer also takes place in related covalently tethered systems, such as **r.2**.

The kinetics of acrylate polymerization using this type of initiators have also been studied [172, 177]; in addition to allowing comparison of the effectiveness of different initiators, the kinetic behavior also provides support for a 2P mechanism for initiation with excitation in the visible and near-infrared range. The rate of a radical polymerization, R_p, is given by [178]:

$$R_p = k_p[M]\sqrt{\frac{R_i}{2k_t}},$$ (7)

where k_p and k_t are the polymerization and termination rate constants, which are dependent on the nature of the monomer, [M] is the monomer concentration, and R_i is the rate of initiation. If the initiation is governed by 2PA, R_i is given by [177]:

$$R_{i,2PA} = \frac{1}{2}\Phi_i\delta N_g\left(\frac{I}{h\nu}\right)^2,$$ (8)

where Φ_i is the quantum efficiency of initiation, $h\nu$ is the photon energy, δ is the two-photon cross section at the excitation wavelength, N_g is the concentration of the two-photon chromophore (molecules/volume), and I is the incident intensity (the electron-transfer rate-constant for chromophores such as **q.2**, where electron-transfer is apparently the first step, is included in Φ_i). From Eqs. 7 and 8, it is clear that the rate of polymerization should depend linearly on the incident intensity for two-photon initiation. In contrast, for the case of one-photon initiation, the initiation rate is [178]:

$$R_{i,1PA} = \Phi_i\sigma N_g\frac{I}{h\nu},$$ (9)

where σ is the 1PA cross section at the excitation wavelength and, thus, R_p in this case depends on the square root of the intensity. Figure 16 shows the R_p (as measured from the volume of polymerized resin as a function of exposure conditions) for SR9008 initiated by compound **q.2** under excitation at 600 nm (top panel), which is the 2PA maximum of this compound (see Table 1) and under excitation at 355 nm (bottom panel), within its 1PA band; the intensity dependence at the two wavelengths is consistent with two- and one-photon initiation, respectively.

The most effective wavelength for 2P-microfabrication in acrylates depends on the 2PA spectra of the chromophores used as initiators, as the initiation rate is proportional to the 2PA cross section (see Eq. 8). In the case of polymerization of SR9008 by **q.2**, it was shown that R_p has approximately

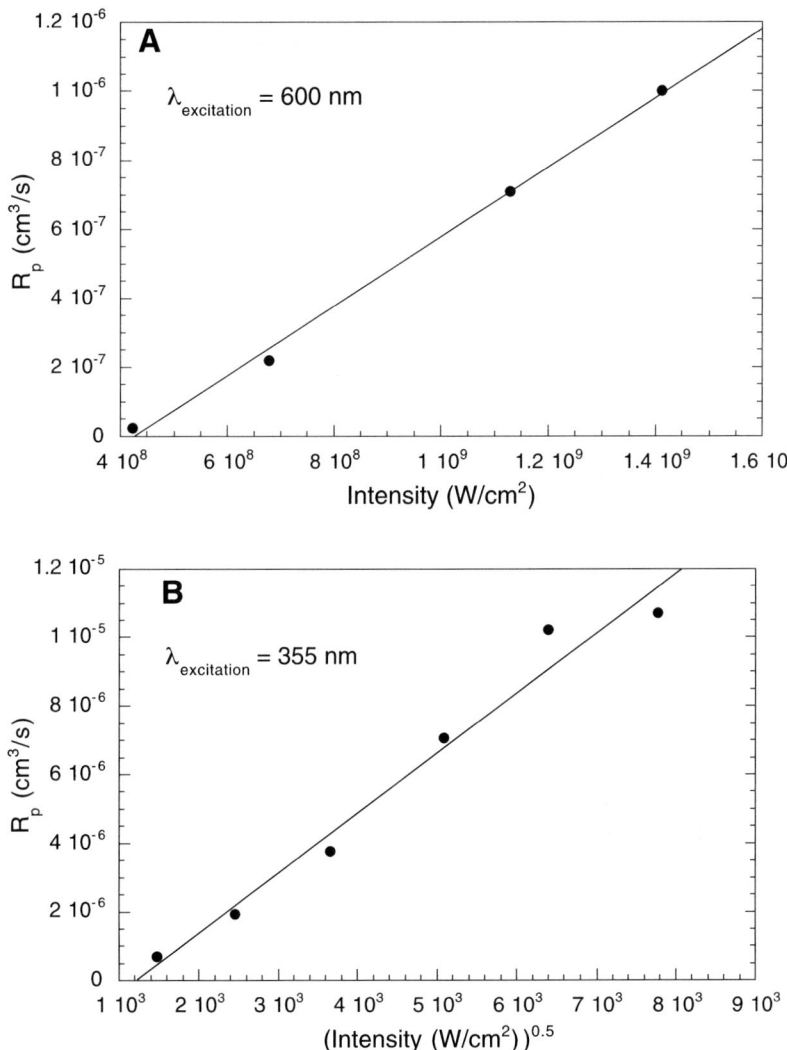

Fig. 16 Rate of polymerization, R_p, of SR9008 initiated by **q.2**: (*top*) as a function of I at 600 nm (5 ns pulses, 20 Hz) and (*bottom*) as a function of $I^{0.5}$ at 355 nm (8 ns pulses, 10 Hz). Reproduced from [172] with permission of the Materials Research Society

the same wavelength dependence as δ [177]. Compounds **q.2**, **r.1**, and **q.11** have 2PA maxima of 600, 730, and 775 nm, respectively, so this class of materials can be used with a wide range of laser sources. More recently, **r.3** (Fig. 13) has also been studied; this chromophore shows a peak cross section of ca. 200 GM at ca. 520 nm [179]. Since the diffraction-limited focal volume is proportional to the third power of the excitation wavelength, decreasing the excitation wavelength can lead to smaller focal volumes, which

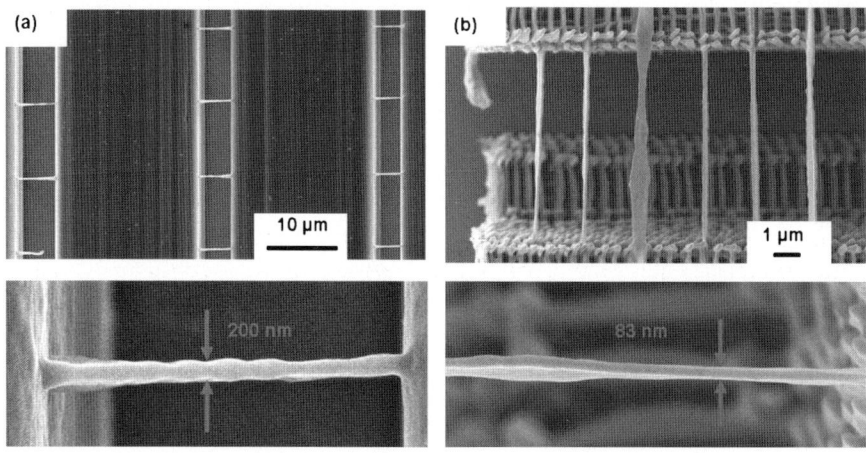

Fig. 17 SEM images of structures fabricated in a triacrylate resin at threshold powers using **a** 730 nm excitation of **r.1** and **b** 520 nm excitation of **r.3**. The *lower panels* show the dimensions, transversal to the beam propagation direction, of single lines of the structures for each excitation wavelength. Reproduced with permission from [179]. © 2007, Optical Society of America

can improve 3D resolution and enable the fabrication of structures with smaller feature sizes. The size of the polymerized features also critically depends on the exposure conditions (such as the dwell time, in addition to the beam intensity) [173, 180, 181]. Figure 17 compares the width of lines between supporting structures written in triacrylate resins using 2P-excitation of **r.1** at 730 nm and of **r.3** at 520 nm (the pulse duration was ∼100 fs in both cases), and demonstrates the greater spatial resolution (∼80 nm) achievable with shorter wavelength excitation than with excitation at 730 nm (∼200 nm)[21] [179]. Even smaller features (down to 60 nm) were written using **r.3** with further optimization of writing conditions. 2P-induced polymerization of acrylate resins has also been demonstrated at 532 nm using a substituted fluorene and a 0.5 ns pulsed laser, and lateral feature sizes of about 160 nm were obtained [182].

On the other hand, 2P-radical polymerization at longer wavelengths (up to 1064 nm) has recently been facilitated by developing extended vinylogues of Michler's ketone (4,4′-bis(N,N-dimethylamino)benzophenone), such as **r.4–r.6** (Fig. 13) [183, 184]; these exhibit peak cross sections from 200–325 GM at 800–950 nm, with appreciable 2P cross section extending to ca. 1100 nm. Waveguides and optical circuitry have been written in acrylate materials using **r.6** [185].

---

[21] The height of the polymerized lines, which is the feature size along the beam propagation direction, was found to be five to seven times larger than their width under the same conditions.

Belfield and co-workers have also fabricated structures from acrylate materials (SR349 from Sartomer) with 775 nm/150 fs laser irradiation using a mixture of c.6 and c.7 (Fig. 13) as initiator system [186]. The 5,7-diiodo-3-butoxy-6-fluorone, c.6, acts as a two-photon chromophore and, in analogy to the previously investigated 1P mechanism for this system [187], photoexcited c.6 in the triplet state is then thought to accept an electron from the amine donor c.7 [175, 186]. Subsequent proton transfer from the amine radical cation methyl group to the fluorone radical anion gives a radical of the form $[ArMeNCH_2]^{\bullet}$ $\{Ar = 2,6{-}^iPr_2C_6H_3\}$, which can then add to an acrylate monomer unit and initiate the polymerization. Belfield's work is described in more detail in another chapter of this book. The 2P radical polymerization of acrylates at 800 nm using another two-component sensitizer, rose Bengal (c.8, Fig. 13)/triethanolamine, is believed to be initiated in a similar way, with electron-transfer from the amine to the triplet state of the dye being the first step [188]. Rose Bengal has also been used for microfabrication using biomolecules (bovine serum albumin and fibrinogen); however, this reaction is inhibited by triethanolamine and proceeds by a different mechanism (see Sect. 5.6) [188–190].

In addition to multi-photon patterning using $^1O_2$-sensitizing dyes such as rose Bengal (discussed in Sect. 5.6), biomolecules have been patterned using multi-photon-induced radical chemistry. Three-photon excitation at 780 nm of a dimeric benzophenone derivative [191], c.9 (Fig. 13), has been used to fabricate model tissue-engineering scaffolds directly from collagens [190]. The ketyl diradical formed by excitation of the benzophenone moiety is believed to undergo a radical reaction with a protein; the presence of two benzophenone groups in the molecule leads to crosslinking. Unlike rose Bengal (see Sect. 5.6), this chromophore can be used in acidic conditions, which is important due to the solubility of collagens in such environments. These proteins retained some of their activity and were shown to display highly specific cell adhesion. Flavin adenine dinucleotide, a less cytotoxic alternative to rose Bengal and benzophenone derivatives and, thus, more suitable for use in vivo, has been used as a 2PA chromophore for patterning of various proteins with laser excitation at 750–790 nm [192–194]. The initiation mechanism is not clear, since O_2^{-}, $^1O_2$, and flavin radicals have all been observed as photoproducts of flavins, but radical processes are likely to play a role in the crosslinking [195, 196].

Another area of potential application for 2P-induced radical reactions is in the generation of high-density 3D optical memories; these could offer large increases in data capacity relative to traditional 2D magnetic or optical media. One such scheme has been demonstrated based on 2PA in chromophore r.2 (Fig. 13); "bits" were written using 600 nm/4 ps laser irradiation in a film consisting of r.2, a triacrylate, and an inert polymer binder [21]. The unwritten areas are nonfluorescent due to intramolecular quenching of the fluorescence of the chromophore by the acrylate side groups. However, when these acrylate groups are incorporated in the polymer chains, they are not as readily reduced as acrylate monomers, and they do not quench the fluorescence of the chromophore as effectively; therefore, the fluorescence of the bis(dialkylamino)stilbene portion is "turned on" in the written areas by radical polymerization of the acrylates [21, 197]. Many other schemes for 2P-induced memory have been demonstrated [1, 2, 17, 18, 198–208], several of which involved the use of high-δ chromophores [201, 204, 206, 208]. Besides the radical chemistry highlighted in the example at the beginning of the paragraph, different chemistries have been used; several studies have been based on 2P-excitation of photochromic molecules [1, 199], although photochromic chromophores with cross sections comparable to chromophores discussed in Sect. 4 have not yet been developed, while another involves 2P excitation of a photoacid [17, 206]. In addition, alternative "read-out" mechanisms include "turning off" fluorescence by photobleaching of a fluorescence 2PA chromophore [204] and mechanisms in which fluorescence is not involved at all, but in which information is stored and retrieved exploiting refractive-index changes in the recording medium [2].

5.3
Two-Photon Acid Generation

Much of the multi-photon microfabrication reported to date is based on radical polymerization, as discussed in the previous section. However, cationic polymerization using two-photon excitation of photoacid generators has also been demonstrated; one potential advantage is the low shrinkage that occurs on cationic polymerization of some classes of monomers (e.g., epoxides) and which facilitates the high fidelity writing of structures [7]. In addition, cationic polymerization is generally less sensitive than radical polymerization to the presence of oxygen. Photoacids also make possible the use of two-photon microfabrication in positive-tone resists, i.e., materials whose solubility increases on irradiation.

As with radical polymerization, some studies have used well-established 1P-initiators under 2P excitation conditions. For example, the commercially available photoacid generator **p.1** (Fig. 18), which has a peak cross section of only 16 GM at a photon wavelength of 530 nm [169], has been used to fabricate

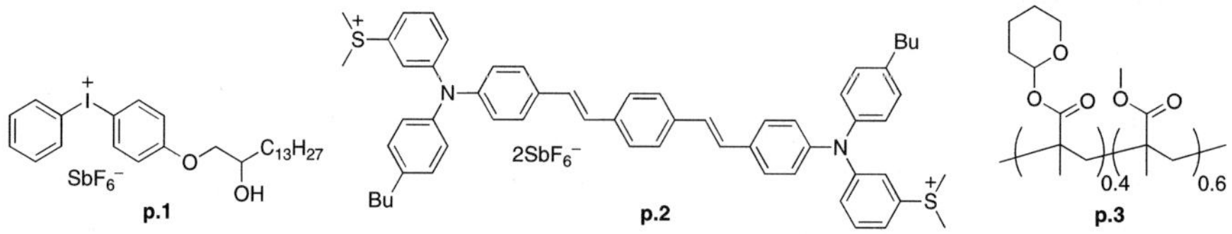

Fig. 18 Structures discussed in the context of 2P-photoacid chemistry

microstructures of lines in commercial epoxide resins (K126 from Sartomer) under 775 nm/150 fs excitation [175]. The dye c.4 (Fig. 13) has a 2PA peak at 760 nm, red-shifted relative to that of p.1 [169], and has been used to sensitize p.1 (presumably through 2P-induced electron-transfer to p.1) for the polymerization of the epoxide K126 at 710 nm/150 fs; while this epoxide resin could be polymerized with a wide range of powers between the polymerization and damage thresholds, the polymerization rate was limited by the still low cross section of c.4 [209]. In the same study, it was shown that the intensity dependence of R_p for cationic polymerization of K126 is different for excitation at 710 nm (2PA case) and at 365 nm (1PA case), as expected. Also this dependence is different for cationic and radical polymerization at the same order of absorption process, as Eq. 7 is not directly applicable to the former case. The initiator system of coumarin derivative, c.5 (Fig. 13), and $[Ph_2I]^+[PF_6]^-$, previously mentioned in Sect. 5.2 in the context of radical polymerization, has also been used to initiate cationic polymerizations [210]. It was also mentioned that optical memory systems based on 2PA-write and/or 2PA-read schemes using conventional photoacids have been reported [17, 206].

Saeva et al. have extensively studied 1P photoacid generation by aryl dialkyl sulfonium salts; variation of the aryl group or attachment to other chromophores can be used to tune the wavelength at which these chromophores undergo cleavage [211]. The proposed mechanism of photocleavage [212, 213] is that a $\pi-\pi^*$ excited state is formed on photoexcitation; an electron is then transferred from the π^* orbital to a S–R (R = alkyl) σ^* orbital, leading to homolytic cleavage of the S–R bond. Attack of the resulting alkyl radical on the aryl group gives a cationic species which can then lose a proton. The same basic design principle has been used to develop a new photoacid generator, p.2 (Fig. 18), with a large 2PA cross section and a high quantum yield for acid generation (0.50 ± 0.05) [214, 215]. The 2PA spectrum of p.2 is shown in Fig. 19 (top) and exhibits strong (> 100 GM) 2PA at photon wavelengths from 705 to 850 nm with the highest cross section measured being 690 GM at ca. 710 nm, similar to findings for other bis(aminostyryl)benzene chromophores (see Sect. 4.2). The acid-yield efficiency spectrum, also shown in Fig. 19 (top) and representing the relative concentration of acid produced by irradiation of a solution of p.2 in acetonitrile for 30 min at different excitation wavelengths, is similar to the 2PA spectrum of the material, indicating a direct connection between the 2P excitation of p.2 and the acid generation. In addition, the acid concentration after irradiation at 745 nm increases quadratically with excitation power (Fig. 19 (bottom)), as expected for a photochemical process activated by 2PA [214, 215]. The 2P-photoacid p.2, has been used for the 2P-induced cationic polymerization of crosslinkable epoxide-based photoresists, including SU-8, under ns and fs laser irradiation. The polymerization threshold at 710 and 760 nm was at least an order of magnitude lower for p.2 than for p.1 or for the p.1/c.4 system in the resin Araldite CY179MA (from Cyba). Patterning of positive-tone resists based on the polymer p.3 (Fig. 18) with the photoacid

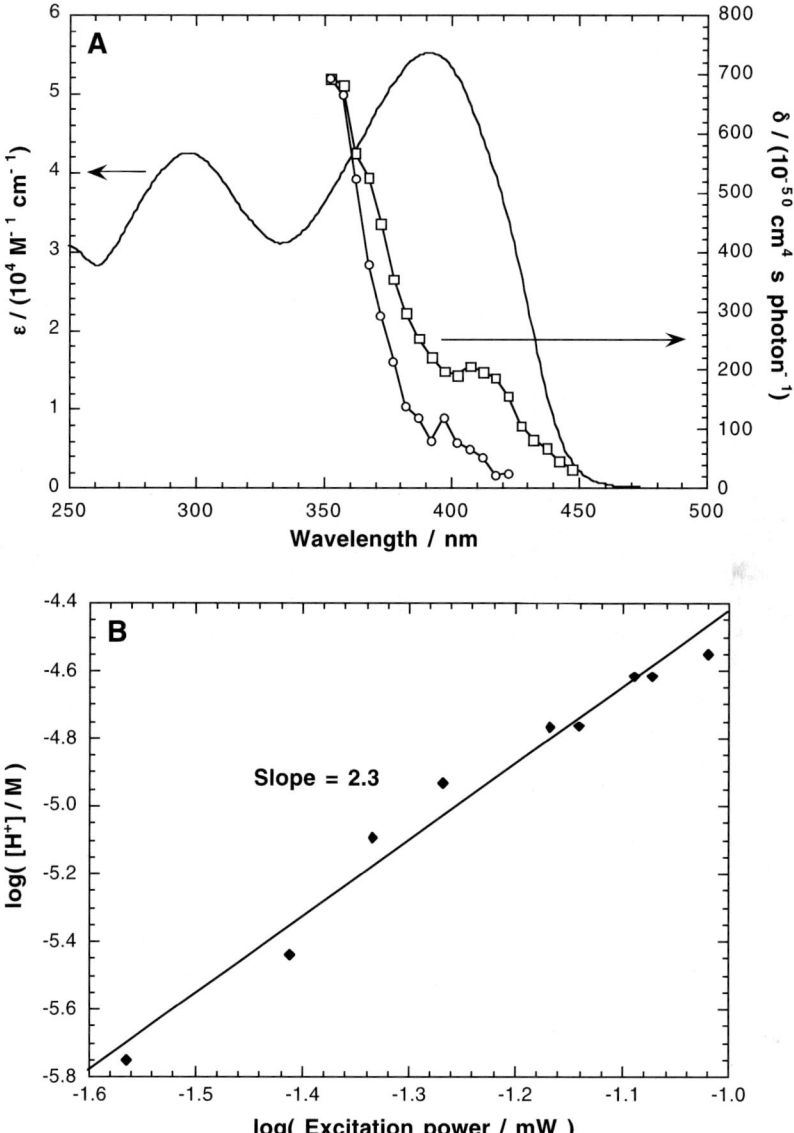

Fig. 19 *Top*: One-photon absorption (*solid line*), two-photon excitation (□), and relative acid-yield efficiency (○) spectra of **p.2** plotted versus *transition* wavelength. *Bottom*: Logarithmic plot of [H⁺] against power of the excitation beam at 745 nm/80 fs pulses; the best-fit line has a slope of 2.3, indicating an approximate quadratic increase of acid yield with excitation intensity. Reproduced from [214] with permission from AAAS

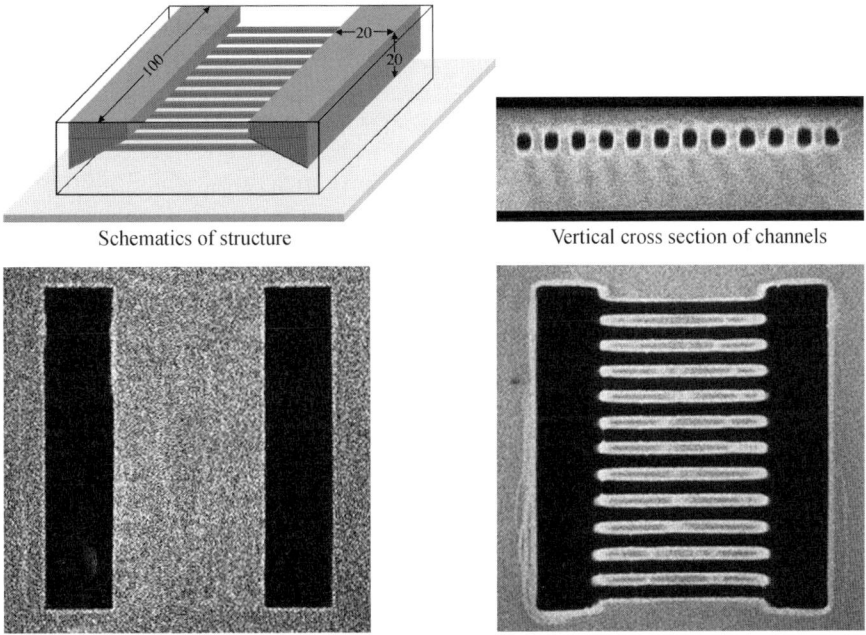

Schematics of structure | Vertical cross section of channels

Film surface | 10 μm below surface

Fig. 20 *Top left*: Schematic of a buried channels structure fabricated using photoacid **p.2** and positive-resist **p.3** (745 nm/80 fs). *Top right*: Two-photon fluorescence micrograph of a vertical cross section, perpendicular to the channels (the channel-to-channel spacing is 8 μm); *Bottom*: Two-photon micrographs of the structure at the surface and at a depth of 10 μm (the length of the channels is 50 μm). The sections where the polymer has been removed appear dark in the images. Adapted from [214]. Reproduced with permission from AAAS

p.2 has also been demonstrated at 745 nm. The exposed areas are soluble in aqueous $[Me_4N]^+[OH]^-$ due to the photoacid-induced cleavage of the tetrahydropyranyl esters to unmask carboxylic acids. An example of channel structure "excavated" in this fashion is shown in Fig. 20 [214].

5.4
Two-Photon Patterning of Inorganic and Hybrid Materials

The fabrication of structures in a variety of materials besides polymers, including metals, semiconductors, and inorganic–organic hybrids, is desirable for a variety of applications. For example, in photonic crystals, complete band gaps can be obtained only if the refractive index of the material is above a minimum value, usually larger than achievable with organic polymers [216]. A variety of strategies have been applied to patterning materials in each of these classes by multiphoton processes.

One strategy is to fabricate a template structure using polymeric material (thus, using the same chemistry as described in Sects. 5.2 and 5.3) and back-fill or coat this structure with inorganic materials. For example, surface modification, followed by electroless deposition of Ag [217–219] or Cu [220], or by chemical reduction of Au^{III} solutions by surface functionalities [220], has been used to obtain metallized structures, while infiltration of polymeric photonic bandgap-type structures with $Ti(O^iPr)_4$ solution, followed by hydrolysis and calcination, has been used to obtain highly refractive inverted TiO_2 structures [221]. Au has also been deposited onto multiphoton-patterned matrices of biomaterials [194].

Direct multi-photon patterning of inorganic materials has been achieved in a variety of ways. As_2S_3 has been patterned at 800 nm, the key process being the polymerization of As_4S_6 molecules [222]. Metallic Fe, Ag, and Au have all been deposited from solutions of their compounds or ions [223–228], although there are a number of technical difficulties which limit the scope of solution methods [9]. Metallic Ag structures have been created at 800 nm through a three-step photographic-type approach based on multiphoton creation of a latent Ag nanoparticle image from reduction of aqueous $AgNO_3$ within a silica matrix, followed by additional fixing and developing steps [229]. Metallic Ag has been patterned in two-dimensions in Ag-nanoparticle-containing films obtained from mixing $AgNO_3$ with poly(vinylpyrrolidone) [230], while metallic Au patterns have been obtained from $HAuCl_4$/poly(vinylalcohol) films [231]. For some of these systems the nature of the multiphoton process is not known; for example, for the Ag deposition in silica matrices, the authors suggested either multiphoton excitation of Ag^+ or multiphoton ionization of water could be the key nonlinear optical process [229]. Also, in some cases, high laser intensities are required, due to the absence of a strongly multiphoton-absorbing chromophore.

The use of organic dyes with high 2PA cross section in the writing of inorganic structures has been much more limited; however, dyes i.1 and i.2 (Fig. 21) have been used to write wires of group-10 metals, especially Ag. In this case, donor-substituted dyes, such as q.1, are unsuitable, since the ground state molecule, as well as the excited state, is capable of reducing Ag^+. However, acceptor-substituted quadrupolar dyes, such as i.1 and i.2,

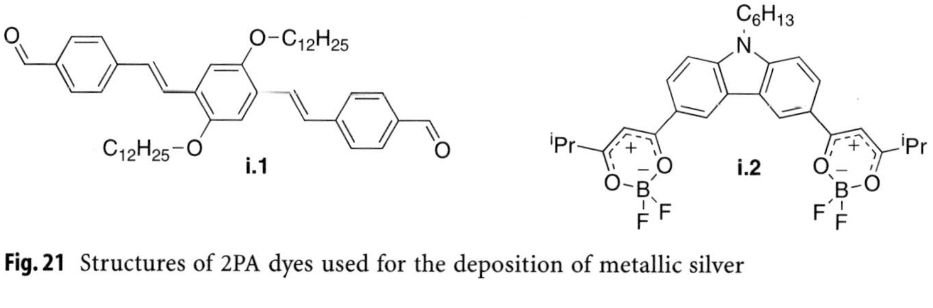

Fig. 21 Structures of 2PA dyes used for the deposition of metallic silver

which are much less easily oxidized than species such as **q.1** and have peak cross sections of 360 GM at 700 nm and ca. 530 GM at ca. 710 nm, respectively, have been used for 2P writing of Ag lines from a resin composed of 2PA dye, thiol-coated Ag nanoparticles, AgBF$_4$, poly(N-vinylcarbazole), and N-ethylcarbazole at 730–800 nm [232, 233].

Organic-silica hybrid materials have been used for multi-photon microfabrication. These include the acrylate-functionalized oligosiloxanes known as ORMOCERs, which have been polymerized by radical processes using conventional 1P radical iniatitors, such as **c.2** [221, 234]. Commercial poly(dimethylsiloxane)-based resists containing vinyl and Si–H functionalities use two different 2PA-induced processes: hydrosilylation catalyzed by the photodecomposition products of Cp′PtMe$_3$ (Cp′ = η^5-methylcyclopentadienyl) and radical initiation by **c.4** (Fig. 13) [235]. The former process was complicated by thermally-induced polymerization.

5.5
Two-Photon Deprotection

The 2P-induced photorelease of protons has been discussed in Sect. 5.3; clearly, the 2P-induced photorelease of other functional groups might facilitate the use of different chemistries and materials in microfabrication. Moreover, the photorelease or deprotection of functional molecules is a subject of considerable interest in biology and medicine; use of 2PA makes possible this release in highly localized volumes (ca. 1 μm^3, which is about the size of a typical bacterium cell, and is smaller than the size of a typical mammalian cell or neuron). For biological and medical applications, it is desirable to excite chromophores in the 700–1100 nm window between the electronic absorption of various biological molecules and absorption by vibrational overtones of water. Light scattering is also reduced in this wavelength range relative to the visible.

While 2P-induced photorelease of a wide variety of functional groups has been demonstrated, systems studied to date have generally used conventional 1P deprotection systems with low near-infrared 2PA cross sections [236–239],

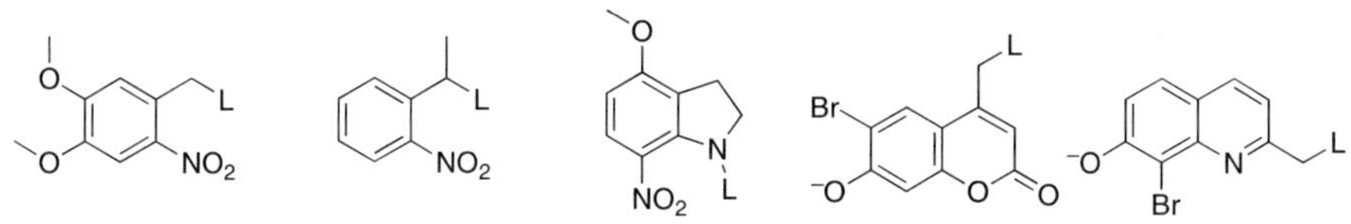

dp.1	dp.2	dp.3	dp.4	dp.5
0.03 GM at 740 nm	0.019 GM at 640 nm	0.06 GM at 720 nm	0.72 GM at 740 nm	0.59 GM at 740 nm

Fig. 22 Some photoremovable protecting groups for which near-infrared two-photon uncaging cross sections have been measured at the wavelength specified [236–239]. In all cases, L represents the protected group

although it has been suggested that 2P-uncaging cross sections as small as 0.1 GM are biologically useful [236]. Some examples, along with their uncaging cross sections, i.e., the product of the 2PA cross section and quantum yield for the photodeprotection reaction, at near-infrared wavelengths are shown in Fig. 22. Coumaryl protecting groups (**dp.4** and similar structures), have the best uncaging cross sections of the widely studied examples and have been used for the photorelease of a wide variety of molecules including glutamates [236], DNA and RNA [240], diols [241, 242], alcohols [243], cyclic nucleotide monophosphates [244], ketones and aldehydes [245], an inhibitor of nitric oxide synthase [246, 247], carboxylates [242], and phosphates [242]. A challenge for future work will be to develop photodeprotection systems with both high near-infrared cross sections (based, for example, on the molecular structures discussed in Sect. 4) and efficient photodeprotection quantum yields.

5.6
Singlet-Oxygen Generation

Another area of biomedical interest in 2PA is in 2P photodynamic therapy, in which 2PA is used to generate cytotoxic singlet O_2 to destroy cancer cells [248]. The lowest singlet state of molecular oxygen lies 94.2 kJ mol^{-1} (ca. 1 eV) above the triplet ground state. Singlet O_2 is typically photogenerated by energy transfer from the lowest triplet excited state (T_1) of a dye to the triplet ground state (T_0) of O_2, the dye relaxing back to its singlet ground state (S_0). Thus, chromophores of interest should have a S_0–T_1 separation of greater than 94.2 kJ mol^{-1} and should form the T_1 state in reasonable yield on photoexcitation, i.e., should exhibit an intersystem crossing rate that is competitive with other processes that can deactivate S_1 (such as fluorescence and internal conversion). As with biologically targeted 2P-photorelease (Sect. 5.5), the 2PA process should take place in the near-infrared window of tissue transparency.

2P-induced singlet O_2 generation has been demonstrated with several chromophores with high near-infrared 2PA cross sections including **q.5** (Fig. 9), for which the quantum yield of $^1O_2$ generation was found to be 0.11 [249]. Replacement of the cyano groups of **q.5** with bromine atoms leads to an increased $^1O_2$ quantum yield of 0.46 [249, 250], presumably due to the high spin-orbit coupling constant associated with the heavy bromine atom. Subsequently, water-soluble analogues of the bromine-substituted species, more suitable for use in biological media, have been developed [251]. $^1O_2$ production has been demonstrated for a range of other 2P-absorbing chromophore systems, in some cases in vivo. Acceptor-substituted fluorenes [252–254] and porphyrin- and porphycene-based systems [155, 255–258] have been among the most widely studied. In other systems, less direct mechanisms of 2PA $^1O_2$ generation have been employed, in which energy transfer takes place from the excited 2PA chromophore to a second chromophore. The second chromophore then undergoes intersystem crossing to give a triplet which can then transfer energy to $^3O_2$. The 2PA and $^1O_2$-generating chromophores have been kept in proximity by incorporating both chromophores into dendrimers [259], by encapsulation of one chromophore into micelles formed by a polymer of the other [260], or by encapsulation of both components into silica nanoparticles [261].

One-photon studies of the reaction of photoexcited rose Bengal, c.8 (Fig. 13), with proteins indicate that the triplet state of c.8 sensitizes the production of $^1O_2$ [262, 263], which in turn reacts with the proteins, resulting in their crosslinking. This $^1O_2$ chemistry has been utilized for the 2PA patterning of biomaterials using rose Bengal [188–190]. Scaffolds and extracellular matrices of fibrinogen, fibronectin, and concanavalin A have been fabricated; these proteins retained their bioactivity following the crosslinking process [189]. Methylene blue has also been used for the 2PA-induced $^1O_2$ crosslinking of proteins [193, 194].

References

1. Parthenopoulos DA, Rentzepis PM (1989) Science 245:843
2. Strickler JH, Webb WW (1991) Opt Lett 16:1780
3. Strickler JH, Webb WW (1990) Proc SPIE 1398:107
4. Wu ES, Strickler JH, Harrell WR, Webb WW (1992) Proc SPIE 1674:776
5. Denk W, Strickler JH, Webb WW (1990) Science 248:73
6. Strehmel B, Strehmel V (2007) In: Neckers DC, Jenks WS, Wolff T (eds) Advances in Photochemistry, vol 29. Wiley, Hoboken, New Jersey p 111
7. Sun H-B, Kawata S (2004) Adv Polym Sci 170:169
8. Kuebler SM, Rumi M (2004) In: Guenther BD, Steel DG, Bayvel L (eds) Encyclopedia of Modern Optics, vol 3. Elsevier, Oxford, p 189
9. LaFratta CN, Fourkas JT, Baldacchini T, Farrer RA (2007) Angew Chem Int Ed 46:6238
10. Xu C, Webb WW (1997) Nonlinear and two-photon-induced fluorescence In: Lakowicz J (ed) Topics in fluorescence spectroscopy, vol 5. Plenum Press, New York, p 471
11. So PTC, Dong CY, Masters BR, Berland KM (2000) Annu Rev Biomed Eng 2:399
12. Göppert-Mayer M (1931) Ann Phys (Leipzig) 9:273
13. Peticolas WL (1967) Annu Rev Phys Chem 18:233
14. McClain WM (1974) Acc Chem Res 7:129
15. Lakowicz JR (1999) Principles of fluorescence spectroscopy. Kluver Academic/Plenum Press, New York
16. McClain WM (1971) J Chem Phys 55:2789
17. Dvornikov AS, Rentzepis PM (1997) Opt Commun 136:1
18. Pudavar HE, Joshi MP, Prasad PN, Reinhardt BA (1999) Appl Phys Lett 74:1338
19. Maruo S, Ikuta K (1999) Transducers '99. Sendai, Japan, p. 1232
20. Maruo S, Nakamura O, Kawata S (1997) Opt Lett 22:132
21. Cumpston BH, Ananthavel SP, Barlow S, Dyer DL, Ehrlich JE, Erskine LL, Heikal AA, Kuebler SM, Lee I-YS, McCord-Maughon D, Qin J, Röckel H, Rumi M, Wu X-L, Marder SR, Perry JW (1999) Nature 398:51

22. Borisov RA, Dorojkina GN, Koroteev NI, Kozenkov VM, Magnitskii SA, Malakhov DV, Tarasishin AV, Zheltikov AM (1998) Laser Phys 8:1105
23. Witzgall G, Vrijen R, Yablonovitch E, Doan V, Schwartz BJ (1998) Opt Lett 23:1745
24. Hell SW, Lindek S, Stelzer EHK (1994) J Mod Opt 41:675
25. Piston DW, Kirby MS, Cheng H, Lederer WJ, Webb WW (1994) Appl Opt 33:662
26. Denk W, Delaney KR, Gelperin A, Kleinfeld D, Strowbridge BW, Tank DW, Yuste R (1994) J Neurosci Methods 54:151
27. Monson PR, McClain WM (1970) J Chem Phys 53:29
28. Bredikhin VI, Galanin MD, Genkin VN (1974) Sov Phys-Usp 16:299
29. Friedrich DM, McClain WM (1980) Annu Rev Phys Chem 31:559
30. Harris DC, Bertolucci MD (1978) Symmetry and Spectroscopy – An introduction to vibrational and electronic spectroscopy. Dover Publications, Inc., New York
31. Platt JR (1949) J Chem Phys 17:484
32. Murrell JN (1963) The theory of the electronic spectra of organic molecules. Methuen & Co., Ltd., London, pp 96–100
33. Rice JK, Anderson RW (1986) J Phys Chem 90:6793
34. Callis PR, Scott TW, Albrecht AC (1983) J Chem Phys 78:16
35. Rava RP, Goodman L (1982) J Am Chem Soc 104:3815
36. Goodman L, Rava RP (1983) Adv Chem Phys 54:177
37. Johnson PM (1980) Acc Chem Res 13:20
38. Chen CH, McCann MP (1988) J Chem Phys 88:4671
39. Hochstrasser RM, Meredith GR, Trommsdorff HP (1980) J Chem Phys 73:1009
40. Goodman L, Rava RP (1984) Acc Chem Res 17:250
41. Chia L, Goodman L (1982) J Chem Phys 76:4745
42. Rehms AA, Callis PR (1993) Chem Phys Lett 208:276
43. Jones RD, Callis PR (1988) J Appl Phys 64:4301
44. Jones RD, Callis PR (1988) Chem Phys Lett 144:158
45. Mikami N, Ito M (1975) Chem Phys Lett 31:472
46. Dick B, Hohlneicher G (1981) Chem Phys Lett 84:471
47. Fang HL, Gustafson TL, Swofford RL (1983) J Chem Phys 78:1663
48. Rehms AA, Callis PR (1987) Chem Phys Lett 140:83
49. Meech SR, Phillips D, Lee AG (1983) Chem Phys 80:317
50. Anderson BE, Jones RD, Rehms AA, Ilich P, Callis PR (1986) Chem Phys Lett 125:106
51. Drucker RP, McClain WM (1974) J Chem Phys 61:2609
52. Drucker RP, McClain WM (1974) J Chem Phys 61:2616
53. Hochstrasser RM, Sung H-N, Wessel JE (1973) J Chem Phys 58:4694
54. Dick B, Hohlneicher G (1983) Chem Phys Lett 97:324
55. Hudson B, Kohler B (1974) Annu Rev Phys Chem 25:437
56. Hudson BS, Kohler BE, Schulten K (1982) In: Lim EC (ed) Excited states, vol 6. Academic Press, New York, p 1
57. Schulten K, Karplus M (1972) Chem Phys Lett 14:305
58. Stachelek TM, Pazoha TA, McClain WM, Drucker RP (1977) J Chem Phys 66:4540
59. Anderson RJM, Holtom GR, McClain WM (1977) J Chem Phys 66:3832
60. Anderson RJM, Holtom GR, McClain WM (1979) J Chem Phys 70:4310
61. Chen CH, McCann MP (1987) Opt Commun 63:335
62. Swofford RL, McClain WM (1973) J Chem Phys 59:5740
63. Swofford RL, McClain WM (1975) Chem Phys Lett 34:455
64. Birge RR (1983) In: Kliger DS (ed) Ultrasensitive laser spectroscopy. Academic Press, New York, p 109

65. Chapple PB, Staromlynska J, Hermann JA, McKay TJ, McDuff RG (1997) J Nonlinear Opt Phys Mater 6:251
66. Sutherland RL, Rea E, Natarajan LV, Pottenger T, Fleitz PA (1993) J Chem Phys 98:2593
67. Holtom GR, McClain WM (1976) Chem Phys Lett 44:436
68. Fang HL-B, Thrash RJ, Leroi GE (1978) Chem Phys Lett 57:59
69. Fang HLB, Thrash RJ, Leroi GE (1977) J Chem Phys 67:3389
70. Granville MF, Holtom GR, Kohler BE, Christensen RL, D'Amico KL (1979) J Chem Phys 70:593
71. Kohler BE, Terpougov V (1996) J Chem Phys 104:9297
72. Kohler BE, Spangler C, Westerfield C (1988) J Chem Phys 89:5422
73. Galanin MD, Chizhikova ZA (1966) JETP Lett USSR 4:27
74. Bradley DJ, Hutchinson MHR, Koetser H (1972) Proc R Soc Lond Ser A 329:105
75. Li S, She CY (1982) Opt Acta 29:281
76. Wakebe T, Van Keuren E (1999) Jpn J Appl Phys 38:3556
77. Hermann JP, Ducuing J (1972) Opt Commun 6:101
78. Xu C, Webb WW (1996) J Opt Soc Am B 13:481
79. Albota MA, Xu C, Webb WW (1998) Appl Optcs 37:7352
80. Fisher WG, Wachter EA, Lytle FE, Armas M, Seaton C (1998) Appl Spectrosc 52:536
81. Parma L, Omenetto N (1978) Chem Phys Lett 54:541
82. Parma L, Omenetto N (1978) Chem Phys Lett 54:544
83. Galanin MD, Kirsanov BP, Chizhikova ZA (1969) JETP Lett USSR 9:304
84. Bradley DJ, Hutchinson MHR, Koetser H, Morrow T, New GHC, Petty MS (1972) Proc R Soc Lond Ser A 328:97
85. Rowe HE, Li T (1970) IEEE J Quantum Electron QE-6:49
86. Weber HP (1971) IEEE J Quantum Electron QE-7:189
87. Xu C, Guild J, Webb WW, Denk W (1995) Opt Lett 20:2372
88. Sheik-Bahae M, Said AA, Wei T-H, Hagan DJ, Van Stryland EW (1990) IEEE J Quantum Electron 26:760
89. Sutherland RL (1996) Handbook of nonlinear optics. Marcel Dekker, Inc., New York
90. Kennedy SM, Lytle FE (1986) Anal Chem 58:2643
91. Orr BJ, Ward JF (1971) Mol Phys 20:513
92. Butcher PN, Cotter D (1990) The elements of nonlinear optics. Cambridge University Press, Cambridge
93. Kogej T, Beljonne D, Meyers F, Perry JW, Marder SR, Brédas JL (1998) Chem Phys Lett 298:1
94. Rumi M, Ehrlich JE, Heikal AA, Perry JW, Barlow S, Hu Z, McCord-Maughon D, Parker TC, Röckel H, Thayumanavan S, Marder SR, Beljonne D, Brédas J-L (2000) J Am Chem Soc 122:9500
95. Albota M, Beljonne D, Brédas J-L, Ehrlich JE, Fu J-Y, Heikal AA, Hess SE, Kogej T, Levin MD, Marder SR, McCord-Maughon D, Perry JW, Röckel H, Rumi M, Subramaniam G, Webb WW, Wu X-L, Xu C (1998) Science 281:1653
96. Pond SJK, Rumi M, Levin MD, Parker TC, Beljonne D, Day MW, Brédas J-L, Marder SR, Perry JW (2002) J Phys Chem A 106:11470
97. Bartholomew GP, Rumi M, Pond SJK, Perry JW, Tretiak S, Bazan GC (2004) J Am Chem Soc 126:11529
98. Morel Y, Irimia A, Najechalski P, Kervella Y, Stephan O, Baldeck PL, Andraud C (2001) J Chem Phys 114:5391
99. Kim O-K, Lee K-S, Woo HY, Kim K-S, He GS, Swiatkiewicz J, Prasad PN (2000) Chem Mater 12:284

100. Ventelon L, Charier S, Moreaux L, Mertz J, Blanchard-Desce M (2001) Angew Chem Int Ed 40:2098
101. Ventelon L, Moreaux L, Mertz J, Blanchard-Desce M (2002) Synth Met 127:17
102. Yang WJ, Kim DY, Jeong M-Y, Kim HM, Jeon S-J, Cho BR (2003) Chem Commun: 2618
103. Kamada K, Ohta K, Iwase Y, Kondo K (2003) Chem Phys Lett 372:386
104. Oudar JL, Chemla DS (1977) J Chem Phys 66:2664
105. Marder SR, Beratan DN, Cheng L-T (1991) Science 252:103
106. Marder SR, Perry JW, Bourhill G, Gorman CB, Tiemann BG, Mansour K (1993) Science 261:186
107. Marder SR, Gorman CB, Meyers F, Perry JW, Bourhill G, Brédas J-L, Pierce BM (1994) Science 265:632
108. Gorman CB, Marder SR (1993) Proc Natl Acad Sci USA 90:11297
109. Reinhardt BA, Brott LL, Clarson SJ, Dillard AG, Bhatt JC, Kannan R, Yuan L, He GS, Prasad PN (1998) Chem Mater 10:1863
110. Fleitz PA, Brant MC, Sutherland RL, Strohkendl FP, Larsen RJ, Dalton LR (1998) Proc SPIE 3472:91
111. Swiatkiewicz J, Prasad PN, Reinhardt BA (1998) Opt Commun 157:135
112. Audebert P, Kamada K, Matsunaga K, Ohta K (2003) Chem Phys Lett 367:62
113. Antonov L, Kamada K, Ohta K, Kamounah FS (2003) Phys Chem Chem Phys 5:1193
114. Delysse S, Raimond P, Nunzi J-M (1997) Chem Phys 219:341
115. Strehmel B, Sarker AM, Detert H (2003) Chem Phys Chem 4:249
116. Chérioux F, Maillotte H, Dudley JM, Audebert P (2000) Chem Phys Lett 319:669
117. Zyss J (1993) J Chem Phys 98:6583
118. Beljonne D, Wenseleers W, Zojer E, Shuai Z, Vogel H, Pond SJK, Perry JW, Marder SR, Brédas J-L (2002) Adv Funct Mater 12:631
119. Cho BR, Son KH, Lee SH, Song Y-S, Lee Y-K, Jeon S-J, Choi JH, Lee H, Cho M (2001) J Am Chem Soc 123:10039
120. Yang WJ, Kim CH, Jeong M-Y, Lee SK, Piao MJ, Jeon S-J, Cho BR (2004) Chem Mater 16:2783
121. Porrès L, Katan C, Mongin O, Pons T, Mertz J, Blanchard-Desce M (2004) J Mol Struct 704:17
122. Yoo J, Yang SK, Jeong M-Y, Ahn HC, Jeon S-J, Cho BR (2003) Org Lett 5:645
123. Katan C, Terenziani F, Mongin O, Werts MHV, Porrès L, Pons T, Mertz J, Tretiak S, Blanchard-Desce M (2005) J Phys Chem A 109:3024
124. Kuzyk MG (2003) J Chem Phys 119:8327
125. Adronov A, Fréchet JMJ, He GS, Kim K-S, Chung S-J, Swiatkiewicz J, Prasad PN (2000) Chem Mater 12:2838
126. Drobizhev M, Karotki A, Rebane A, Spangler CW (2001) Opt Lett 26:1081

127. Drobizhev M, Karotki A, Dzenis Y, Rebane A, Suo Z, Spangler CW (2003) J Phys Chem B 107:7540
128. Drobizhev M, Rebane A, Suo Z, Spangler CW (2005) J Lumines 111:291
129. Wenseleers W, Stellacci F, Meyer-Friedrichsen T, Mangel T, Bauer CA, Pond SJK, Marder SR, Perry JW (2002) J Phys Chem B 106:6853
130. Cho BR, Piao MJ, Son KH, Lee SH, Yoon SJ, Jeon S-J, Cho M (2002) Chem Eur J 8:3907
131. Najechalski P, Morel Y, Stéphan O, Baldeck PL (2001) Chem Phys Lett 343:44
132. Hohenau A, Cagran C, Kranzelbinder G, Scherf U, Leising G (2001) Adv Mater 13:1303
133. Pfeffer N, Raimond P, Charra F, Nunzi J-M (1993) Chem Phys Lett 201:357

134. Morel Y, Ibanez A, Nguefack C, Andraud C, Collet A, Nicoud J-F, Baldeck PL (2000) Synth Met 115:265
135. Pierce BM (1991) Proc SPIE 1560:148
136. Dirk CW, Cheng L-T, Kuzyk MG (1992) Int J Quantum Chem 43:27
137. Stevenson SH, Donald DS, Meredith GR (1988) Mater Res Soc Symp Proc 109:103
138. Johr T, Werncke W, Pfeiffer M, Lau A, Dähne L (1995) Chem Phys Lett 246:521
139. Werncke W, Pfeiffer M, Johr T, Lau A, Grahn W, Johannes H-H, Dähne L (1997) Chem Phys 216:337
140. Brédas J-L, Adant C, Tackx PCE, Persoons A, Pierce BM (1994) Chem Rev 94:243
141. Andrews JH, Khaydarov JDV, Singer KD, Hull DL, Chuang KC (1995) J Opt Soc Am B 12:2360
142. Mathis KS, Kuzyk MG, Dirk CW, Tan A, Martinez S, Gampos G (1998) J Opt Soc Am B 15:871
143. Feldner A, Scherer D, Welscher M, Vogtmann T, Schwoerer M, Lawrentz U, Laue T, Johannes H-H, Grahn W (2000) Nonlinear Opt 26:99
144. Scherer D, Dörfler R, Feldner A, Vogtmann T, Schwoerer M, Lawrentz U, Grahn W, Lambert C (2002) Chem Phys 279:179
145. Fu J, Padilha LA, Hagan DJ, Van Stryland EW, Przhonska OV, Bondar MV, Slominsky YL, Kachkovski AD (2007) J Opt Soc Am B 24:67
146. Belfield KD, Bondar MV, Hernandez FE, Przhonska OV, Yao S (2006) Chem Phys 320:118
147. Chung S-J, Zheng S, Odani T, Beverina L, Fu J, Padilha LA, Biesso A, Hales JM, Zhan X, Schmidt K, Ye A, Zojer E, Barlow S, Hagan DJ, Van Stryland EW, Yi Y, Shuai Z, Pagani GA, Brédas J-L, Perry JW, Marder SR (2006) J Am Chem Soc 128:14444
148. McWilliams PCM, Hayden GW, Soos ZG (1988) Phys Rev B 38:9777
149. Chandross M, Shimoi Y, Mazumdar S (1997) Chem Phys Lett 280:85
150. Perry JW (1997) In: Nalwa HS, Miyata S (eds) Nonlinear optics of organic molecules and polymers. CRC Press, Boca Raton, p 813
151. Karotki A, Drobizhev M, Kruk M, Spangler C, Nickel E, Mamardashvili N, Rebane A (2003) J Opt Soc Am B 20:321
152. Drobizhev M, Karotki A, Kruk M, Mamardashvili NZ, Rebane A (2002) Chem Phys Lett 361:504
153. Drobizhev M, Makarov NS, Stepanenko Y, Rebane A (2006) J Chem Phys 124:224701
154. Drobizhev M, Stepanenko Y, Dzenis Y, Karotki A, Rebane A, Taylor PN, Anderson HL (2004) J Am Chem Soc 126:15352
155. Drobizhev M, Stepanenko Y, Dzenis Y, Karotki A, Rebane A, Taylor PN, Anderson HL (2005) J Phys Chem B 109:7223
156. Ogawa K, Ohashi A, Kobuke Y, Kamada K, Ohta K (2005) J Phys Chem B 109:22003
157. Screen TEO, Thorne JRG, Denning RG, Bucknall DG, Anderson HL (2002) J Am Chem Soc 124:9712
158. Drobizhev M, Stepanenko Y, Rebane A, Wilson CJ, Screen TEO, Anderson HL (2006) J Am Chem Soc 128:12432
159. McDonagh AM, Humphrey MG, Samoc M, Luther-Davies B (1999) Organometallics 18:5195
160. McDonagh AM, Humphrey MG, Samoc M, Luther-Davies B, Houbrechts S, Wada T, Sasabe H, Persoons A (1999) J Am Chem Soc 121:1405
161. Hurst SK, Humphrey MG, Isoshima T, Wostyn K, Asselberghs I, Clays K, Persoons A, Samoc M, Luther-Davies B (2002) Organometallics 21:2024
162. McKay TJ, Staromlynska J, Wilson P, Davy J (1999) J Appl Phys 85:1337

163. Staromlynska J, McKay TJ, Bolger JA, Davy JR (1998) J Opt Soc Am B 15:1731
164. Samoc M, Morrall JP, Dalton GT, Cifuentes MP, Humphrey MG (2007) Angew Chem Int Ed 46:731
165. Powell CE, Morrall JP, Ward SA, Cifuentes MP, Notaras EGA, Samoc M, Humphrey MG (2004) J Am Chem Soc 126:12234
166. Rogers JE, Slagle JE, Krein DM, Burke AR, Hall BC, Fratini A, McLean DG, Fleitz PA, Cooper TM, Drobizhev M, Makarov NS, Rebane A, Kim K-Y, Farley R, Schanze KS (2007) Inorg Chem 46:6483
167. Kawata S, Sun H-B, Tanaka T, Takada K (2001) Nature 412:697
168. Maruo S, Kawata S (1998) J MEMS 7:411
169. Schafer KJ, Hales JM, Balu M, Belfield KD, Van Stryland EW, Hagan DJ (2004) J Photochem Photobiol A Chem 162:497
170. Li C, Luo L, Wang S, Huang W, Gong Q, Yang Y, Feng S (2001) Chem Phys Lett 340:444
171. Gao F, Yang YY (1999) Chin J Polym Sci 17:589
172. Cumpston BH, Ehrlich JE, Erskine LL, Heikal AA, Hu Z-Y, Lee I-YS, Levin MD, Marder SR, McCord DJ, Perry JW, Röckel H, Rumi M, Wu X-L (1998) Mater Res Soc Symp Proc 488:217
173. Kuebler SM, Rumi M, Watanabe T, Braun K, Cumpston BH, Heikal AA, Erskine LL, Thayumanavan S, Barlow S, Marder SR, Perry JW (2001) J Photopolym Sci Technol 14:657
174. Joshi MP, Pudavar HE, Swiatkiewicz J, Prasad PN, Reianhardt BA (1999) Appl Phys Lett 74:170
175. Belfield KD, Schafer KJ, Liu Y, Liu J, Ren X, Van Stryland EW (2000) J Phys Org Chem 13:837
176. Lu Y, Hasegawa F, Goto T, Ohkuma S, Fukuhara S, Kawazu Y, Totani K, Yamashita T, Watanabe T (2004) J Mater Chem 14:75
177. Kuebler SM, Cumpston BH, Ananthavel S, Barlow S, Ehrlich JE, Erskine LL, Heikal AA, McCord-Maughon D, Qin J, Röckel H, Rumi M, Marder SR, Perry JW (2000) Proc SPIE 3937:97
178. Odian G (1981) Principles of Polymerization. John Wiley & Sons, New York
179. Haske W, Chen VW, Hales JM, W. D, Barlow S, Marder SR, Perry JW (2007) Opt Express 15:3426
180. Tanaka T, Sun H-B, Kawata S (2002) Appl Phys Lett 80:312
181. Sun H-B, Takada K, Kim M-S, Lee K-S, Kawata S (2003) Appl Phys Lett 83:1104
182. Wang I, Bouriau M, Baldeck PL, Martineau C, Andraud C (2002) Opt Lett 27:1348
183. Martineau C, Lemercier G, Andraud C, Wang I, Bouriau M, Baldeck PL (2003) Synth Met 138:353
184. Lemercier G, Martineau C, Mulatier J-C, Wang I, Stéphan O, Baldech P, Andraud C (2006) New J Chem 30:1606
185. Klein S, Barsella A, Leblond H, Bulou H, Fort A, Andruad C, Lemercier G, Mulatier JC, Dorkenoo K (2005) Appl Phys Lett 86:211118
186. Belfield KD, Ren X, Van Stryland EW, Hagan DJ, Dubikovsky V, Miesak EJ (2000) J Am Chem Soc 122:1217
187. Hassoon S, Neckers DC (1995) J Phys Chem 99:9416
188. Pitts JD, Campagnola PJ, Epling GA, Goodman SL (2000) Macromolecules 33:1514
189. Basu S, Campagnola PJ (2004) J Biomed Mater Res 71A: 359
190. Basu S, Cunningham LP, Pins GD, Bush KA, Taboada R, Howell AR, Wang J, Campagnola PJ (2005) Biomacromolecules 6:1465

191. Pitts JD, Howell AR, Taboada R, Banerjee I, Wang J, Goodman SL, Campagnola PJ (2002) Photochem Photobiol 76:135
192. Kaehr B, Allen R, Javier DJ, Currie J, Shear JB (2004) Proc Natl Acad Sci 101:16104
193. Allen R, Nielson R, Wise DD, Shear JB (2005) Anal Chem 77:5089
194. Hill RT, Lyon JL, Allen R, Stevenson KJ, Shear JB (2005) J Am Chem Soc 127:10707
195. Spikes JD, Shen H-R, Kopeceková P, Kopecek J (1999) Photochem Photobiol 70:130
196. Egorov SY, Krasnovsky AA, Bashtanov MY, Mironov EA, Ludnikova TA, Kritsky MS (1999) Biochemistry Mosc 64:1117
197. Dyer DJ, Cumpston BH, McCord-Maughon D, Thayumanavan S, Barlow S, Perry JW, Marder SR (2004) Nonlinear Opt 31:175
198. Dvornikov AS, Cokgor I, McCormick F, Piyaket R, Esener S, Rentzepis PM (1996) Optics Commun 128:205
199. Akimov DA, Fedotov AB, Koroteev NI, Magnitskii SA, Naumov AN, Sidorov-Biryukov DA, Zheltikov AM (1997) Jpn J Appl Phys 36:426
200. Koroteev NI, Krikunov SA, Magnitskii SA, Malakhov DV, Shubin VV (1998) Jpn J Appl Phys 37:2279
201. Prasad PN, Bhawalkar JD, Cheng PC, Pan SJ, Shih A, Kumar ND, Ruland G, Burzynski R (1996) Polym Mater Sci Eng 75:173
202. Xia AD, Wada S, Tashiro H, Huang WH (1999) Proc SPIE 3740:402
203. Yamasaki K, Juodkazis S, Watanabe M, Sun HB, Matsuo S, Misawa H (2000) Appl Phys Lett 76:1000
204. Shen Y, Swiatkiewicz J, Prasad PN, Vaia RA (2001) Opt Commun 200:9
205. Day D, Gu M, Smallridge A (2001) Adv Mater 13:1005
206. Belfield KD, Schafer KJ, Andrasik SY, Yavuz O, Stryland EWV, Hagan DJ, Hales JM (2002) Proc SPIE 4459:281
207. Olson CE, Previte MJR, Fourkas JT (2002) Nat Mater 1:225
208. Polyzos I, Tsigaridas G, Fakis M, Giannetas V, Persephonis P, Mikroyannidis J (2003) Chem Phys Lett 369:264
209. Boiko Y, Costa JM, Wang M, Esener S (2001) Opt Express 8:571
210. Murakami Y, Coenjarts CA, Ober CK (2004) J Photopolym Sci Tech 17:115
211. Saeva FD (1994) Adv Electron Transfer Chem 4:1
212. Saeva FD, Morgan BP (1984) J Am Chem Soc 106:4121
213. Saeva FD, Morgan BP, Luss HR (1985) J Org Chem 50:4360
214. Zhou W, Kuebler SM, Braun KL, Yu T, Cammack JK, Ober CK, Perry JW, Marder SR (2002) Science 296:1106
215. Kuebler SM, Braun KL, Zhou W, Cammack JK, Yu T, Ober CK, Marder SR, Perry JW (2003) J Photochem Photobiol A Chem 158:163
216. Yablonovitch E (1994) J Mod Opt 41:173
217. Formanek F, Takeyasu N, Tanaka T, Chiyoda K, Ishikawa A, Kawata S (2006) Opt Expr 14:800
218. Chen Y-S, Tal A, Kuebler SM (2007) Chem Mater 19:3858
219. Chen Y-S, Tal A, Torrance DB, Kuebler SM (2006) Adv Funct Mater 16:1739
220. Farrer RA, LaFratta CN, Li L, Praino J, Naughton MJ, Saleh BEA, Teich MC, Fourkas JT (2006) J Am Chem Soc 128:1796
221. Serbin J, Ovsianikov A, Chichkov B (2004) Opt Expr 12:5221
222. Wong S, Deubel M, Perez-Willard F, John S, Ozin GA, Wegener M, Freymann V (2006) Adv Mater 18:265
223. Swanson JR, Friend CM, Chabal YJ (1987) J Chem Phys 87:5028
224. Xu X, Steinfeld JI (1990) Appl Surf Sci 45:281
225. Wexler D, Zink JI, Tutt LW, Lunt SR (1993) J Phys Chem 97:13563

226. Kempa T, Farrer RA, Giersig M, Fourkas JT (2006) Plasmonics 1:45
227. Tanaka T, Ishikawa A, Kawata S (2006) Appl Phys Lett 88:081107
228. Ishikawa A, Tanaka T, Kawata S (2006) Appl Phys Lett 89:113102
229. Wu P-W, Cheng W, Martini IB, Dunn B, Schwartz BJ, Yablonovitch E (2000) Adv Mater 12:1438
230. Baldacchini T, Pons A-C, Pons J, LaFratta CN, Fourkas JT, Sun Y, Naughton MJ (2005) Opt Express 13:1275
231. Kaneko K, Sun H-B, Duan X-M, Kawata S (2003) Appl Phys Lett 83:1426
232. Stellacci F, Bauer CA, Meyer-Friedrichsen T, Wenseleers W, Alain V, Kuebler SM, Pond SJK, Zhang Y, Marder SR, Perry JW (2002) Adv Mater 14:194
233. Halik M, Wenseleers W, Grasso C, Stellacci F, Zojer E, Barlow S, Brédas J-L, Perry JW, Marder SR (2003) Chem Commun: 1490
234. Serbin J, Egbert A, Ostendorf A, Chichkov BN, Houbertz R, Domann G, Schulz J, Cronauer C, Fröhlich L, Popall M (2003) Opt Lett 28:301
235. Coenjarts CA, Ober CK (2004) Chem Mater 16:5556
236. Furuta T, Wang SSH, Dantzker JL, Dore TM, Bybee WJ, Callaway EM, Denk W, Tsien RY (1999) Proc Natl Acad Sci 96:1193
237. Kiskin NI, Chillingworth R, McCray JA, Piston D, Ogden D (2002) Eur Biophys J 30:588
238. Matsuzaki M, Ellis-Davies GCR, Nemoto T, Miyashita Y, Iino M, Kasai H (2001) Nat Neurosci 4:1086
239. Fedoryak OD, Dore TM (2002) Org Lett 4:3419
240. Ando H, Furuta T, Tsien RY, Okamoto H (2001) Nat Genetics 28:317
241. Lin W, Lawrence DS (2002) J Org Chem 67:2723
242. Zhu Y, Pavlos CM, Toscano JP, Dore TM (2006) J Am Chem Soc 128:4267
243. Suzuki AZ, Watanabe T, Kawamoto M, Nishiyama K, Yamashita H, Ishii M, Iwamura M, Furuta T (2003) Org Lett 5:4846
244. Furuta T, Takeuchi H, Isozaki M, Takahashi Y, Kanehara M, Sugimoto M, Watanabe T, Noguchi K, Dore TM, Kurahashi T, Iwamura M, Tsien RY (2004) Chem Bio Chem 5:1119
245. Lu M, Fedoryak OD, Moister BR, Dore TM (2003) Org Lett 5:2119
246. Montgomery HJ, Perdikakis BR, Fishlock D, Lajoie GA, Jervis E, Guillemette JG (2002) Bioorg Med Chem Lett 10:1919
247. Perdikakis BR, Montgomery HJ, Abbott GL, Fishlock D, Lajoie GA, Jervis E, Guillemette JG (2005) Bioorg Med Chem Lett. 13:47
248. Bhawalkar JD, Kumar ND, Zhao CF, Prasad PN (1997) J Clin Laser Med Surg 15:201
249. Frederiksen PK, Jørgensen M, Ogilby PR (2001) J Am Chem Soc 123:1215
250. Arnbjerg J, Johnsen M, Frederiksen PK, Braslavsky SE, Ogilby PR (2006) J Phys Chem A 110:7375
251. Nielsen CB, Johnsen M, Arnbjerg J, Pittelkow M, McIlroy SP, Ogilby PR, Jørgensen M (2005) J Org Chem 70:7065
252. Belfield KD, Corredor CC, Morales AR, Dessources MA, Hernandez FE (2006) J Fluoresc 16:105
253. Belfield KD, Bondar MV, Przhonska OV (2006) J Fluoresc 16:111
254. Andrasik SJ, Belfield KD, Bondar MV, Hernandez FE, Morales AR, Przhonska OV, Yao S (2007) Chem Phys Chem 8:399
255. Morone M, Beverina L, Abbotto A, Silvestri F, Collini E, Ferrante C, Bozio R, Pagani GA (2006) Org Lett 8:2719
256. Ogawa K, Hasegawa H, Inaba Y, Kobuke Y, Inouye H, Kanemitsu Y, Kohno E, Hirano T, Ogura S-I, Okura I (2006) J Med Chem 49:2276
257. Dy JT, Ogawa K, Satake A, Ishizumi A, Kobuke Y (2007) Chem Eur J 13:3491

258. Arnbjerg J, Jiménez-Banzo A, Paterson MJ, Nonell S, Borrell JI, Christiansen O, Ogilby PR (2007) J Am Chem Soc 129:5188
259. Oar MA, Serin JM, Dichtel WR, Fréchet JMJ, Ohulchanskyy TY, Prasad PN (2005) Chem Mater 17:2267
260. Chen C-Y, Tian Y, Cheng Y-J, Young AC, Ka J-W, Jen AK-Y (2007) J Am Chem Soc 129:7220
261. Kim S, Ohulchanskyy TY, Pudavar HE, Pandey RK, Prasad PN (2007) J Am Chem Soc 129:2669
262. Neckers DC (1989) J Photochem Photobiol A 47:1
263. Balasubramanian D, Du X, Zigler JSJ (1990) Photochem Photobiol 52:761

Adv Polym Sci (2008) 213: 97–156
DOI 10.1007/12_2007_126
© Springer-Verlag Berlin Heidelberg
Published online: 9 January 2008

Two-photon Absorbing Photonic Materials: From Fundamentals to Applications

Kevin D. Belfield[1] (✉) · Sheng Yao[1] · Mykhailo V. Bondar[2]

[1]Department of Chemistry and CREOL, College of Optics and Photonics,
University of Central Florida, P.O. Box 162366, Orlando, FL 32816-2366, USA
belfield@mail.ucf.edu

[2]Institute of Physics, Prospect Nauki, 46, Kiev-28, Kiev, 03028, Ukraine

Abstract This chapter first reviews the fundamental aspects of the two-photon absorbing materials developed in our group. The design strategies and syntheses of a variety of fluorene-based conjugated molecules as two-photon absorbing materials, along with their photochemical and photophysical properties are described. The methodology of synthesis of these 2PA chromophores designed with various donor or acceptor groups, conjugation lengths and symmetries and the effects of the structure on the 2PA properties are demonstrated. This is followed by presentation of detailed studies of their linear absorption, steady-state and time-resolved fluorescence, fluorescence life time and anisotropy, excited-state absorption, and two-photon absorption measurements. The photostabilities of these chromophores are also investigated due to importance of this parameter in several emerging applications. The last part of the chapter provides a description of the application of these materials in fluorescence imaging, 3D data storage, photodynamic therapy, and 3D microfabrication.

Keywords Two-photon absorption · Fluorescence anisotropy · Fluorine derivatives · Two-photon 3D data storage · Two-photon dynamic therapy · Two-photon 3D microfabrication

Abbreviations

1PE	One-photon excitation
2PA	Two-photon absorption
2PE	Two-photon excitation
2PF	Two-photon fluorescence
2PFM	Two-photon fluorescence microscopy
2D	Two-dimensional
3D	Three-dimensional
A	Acceptor
ACN	Acetonitrile
Ac$_2$O	Acetic anhydride
BSA	Bovine serum albumin
CAD	Computer-aided design
CFP	Combined Fluid Products Co.
D	Donor
DAPI	4′,6′-diamidino-2-phenylindole
DIC	Differential interference contrast
DMF	N,N-dimethylformamide
DMSO	Dimethylsulfoxide
DOL	Degree of labeling
DPBF	1,3-diphenylisobenzofuran
DSU	Disk scanning unit
EFAB	Electrochemical fabrication
ESA	Excited state absorption
FWHM	Full width at half maximum
GB	Gigabytes
GM	Goeppert–Mayer units, 1 GM = 10^{-50} cm^4 · s · photon^{-1}
HOAc	Acetic acid
HP	Hematophorphyrin
IR	Infrared
KI	Potassium iodide
KOH	Potassium hydroxygen
NaOAc	Sodium acetate
NBS	N-bromosuccinimide
Nd:YAG	Neodymium Yttrium Aluminum Garnet
OLED	Organic light-emitting diode
PAG	Photoacid generator
PDT	Photodynamic therapy
PMMA-co-VBP	Poly[methylmethacrylate-co-(diethylvinylbenzylphosphonate)]
PMT	Photomultiplier

PS   Photosensitizer
PTI   Photon Technology International
QY   Quantum yield
SOS   Sum-over-state
THF   Tetrahydrofuran
TRES   Time-resolved emission spectra
UV   Ultraviolet
WLC   White-light continuum
WORM   Write once read many

1
Introduction

The theory of the simultaneous absorption of two photons was first developed by Goeppert–Mayer in 1931 [1], but remained mainly an intellectual curiosity until the advent of the pulsed laser provided very high-intensity light. For simplicity, two-photon absorption (2PA) can be conceptualized from a semi-classical perspective [1]. In the 2PA process, molecules exposed to high intensity light can undergo near simultaneous absorption of two photons mediated by a so-called 'virtual state' which has no classical analog. The combined energy of the two photons accesses a stable excited state of the molecule. If the two photons are of the same energy or wavelength, the process is referred to as degenerate 2PA. On the other hand, if the two photons are of different energy or wavelength, the process is non-degenerate 2PA.

As light passes through a molecule, the virtual state may form, persisting for a very short duration on the order of a few femtoseconds. 2PA can result if a second photon arrives before decay of this virtual state, with the probability of 2PA scaling with the square of the light intensity. This process is generally termed simultaneous two-photon absorption. Two-photon absorption thus involves the *concerted* interaction of both photons that combine their energies to produce an electronic excitation analogous to that conventionally caused by a single photon of a correspondingly shorter wavelength. Unlike single-photon absorption, whose probability is linearly proportional to the incident intensity, the 2PA process depends on both a spatial and temporal overlap of the incident photons and takes on a quadratic (non-linear) dependence on the incident intensity, resulting in highly localized photoexcitation with a focused beam (Fig. 1).

Due to the need to use expensive and complicated laser pump systems, early development occurred primarily by physicists. Two-photon absorption properties of existing materials were extensively studied in this stage. In early 1990s, some potential applications with this technique were demonstrated, such as two-photon fluorescence (2PF) 3D data storage [2] and 2PF microscopy [3]. However, the 2PA efficiency of the materials is the bottleneck for these applications, and this has triggered the effort to search for new materials with higher

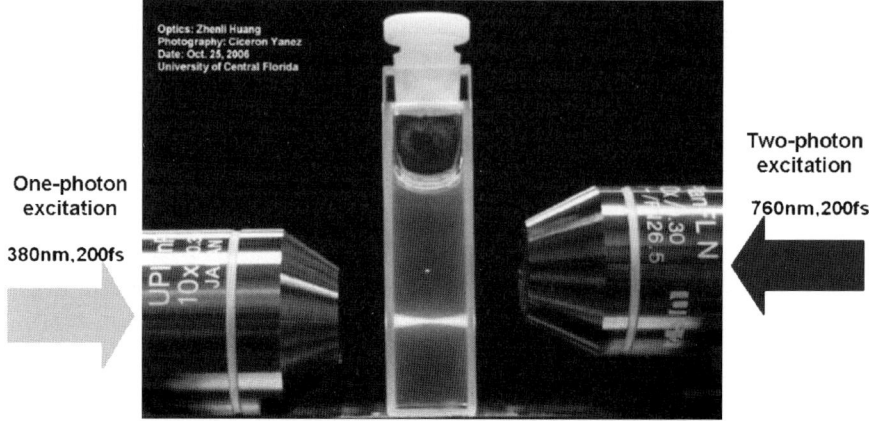

Fig. 1 Demonstration of the spatial selectivity of one-photon excitation (*left*) vs. two-photon excitation (*right*) in a fluorescein solution

2PA efficiency. Breakthrough work conducted by Prasad [4–6], Marder [7–9] and others then opened the door for new design of better 2PA materials. Since then, 2PA remains a very active field both in chemistry and physics. The understanding of fundamental aspects such as relationships between molecular structure and nonlinear absorption properties and two-photon induced photodynamics have greatly advanced the field and provided the tenets for practical applications for this technology in several areas.

The 2PA research in our group began in 1996, involving a number of facets, from fundamental to applications. This review summarizes about 10 years of research efforts on 2PA materials and applications in our group. The first section includes the design and synthesis of novel 2PA chromophores and the chemistry involved in their preparation. The second section is concerned with the linear and nonlinear optical properties of these chromophores as well as their photodynamic behavior. In the last section, further optimization of the chromophores for several intriguing applications is discussed, as are the principles of these applications.

2
Materials Design and Synthesis

2.1
Structures of 2PA Fluorene Derivatives

The materials studied for 2PA properties involve a number of types of conjugated structures with donor and acceptor groups from stryl to cyanine dyes. Our group is particularly interested in conjugated fluorene derivatives. Fluorene derivatives have been used in conducting polymers, more recently in

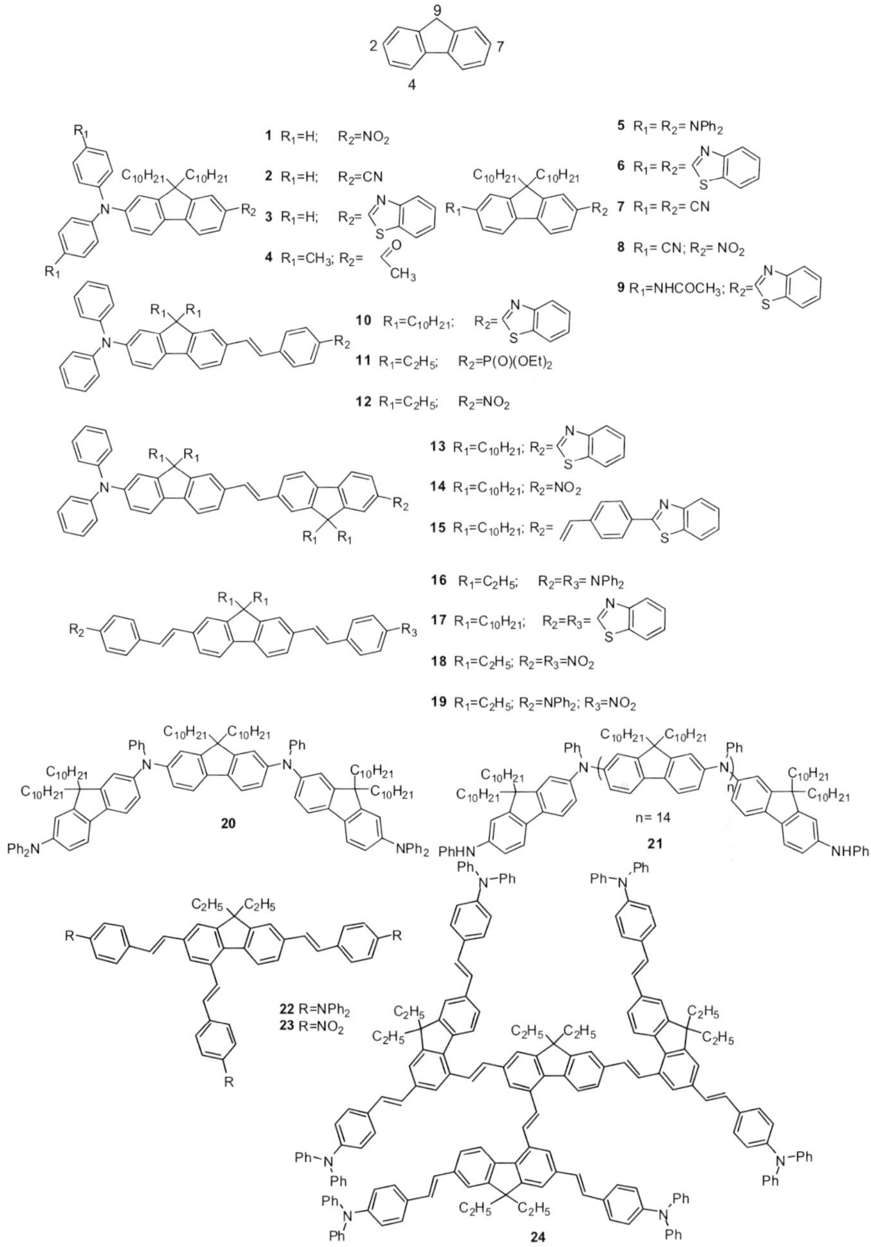

Fig. 2 The structures of fluorene derivatives designed and synthesized for studying molecular structure-2PA property relationships

OLEDs as luminescent sources, especially as blue light emitters characterized by their high fluorescence quantum yields.

As shown in Fig. 2, the biphenyl unit locked into the fluorenyl ring provides a rigid, flat system. As a consequence, the structure provides increased π molecular orbital overlap between the rings giving greater electron delocalization, which serves as a thermally and photochemically stable π-conjugated analogue of the 4,4'-disubstituted biphenyl derivatives. Importantly, fluorene can be readily functionalized in the 2, 4, 7 and/or 9-positions. The reactive 2, 4, and 7 positions have been used to extend the conjugation length, crucial to achieve high 2PA nonlinearity. The reactive 9-position facilitates the introduction of solubilizing groups such as alkyl chains, or in the case of biological probes, more hydrophilic groups such as ethylene-oxy moieties. Our research showed that the various substituents at the 9 position have no effect on the electrooptic properties of the conjugation system, facilitating the modification of the molecules to meet requirements of particular applications.

Since first reported by Reinhardt's group [6, 10, 11], fluorene derivatives have been extensively studied as 2PA materials. The representative structures prepared in our group are shown in Fig. 2. Four factors were systematically varied according to molecular symmetry, electronic character, π-conjugation length and geometry. Linear structures are the basis of the understanding of the 2PA structure-properties relationships, therefore, dyes 1–19 were designed whose structures are shown in Fig. 2. Within these basic structures, they can be classified as symmetric D-π-D or A-π-A (D = donor; A = acceptor) and unsymmetric D-π-A architectures. The molecular symmetry has significant influence on the λ_{max}^{2PA}, which is governed by the different selection rules of 2PA relative to single-photon absorption.

It is also evident that from 1–19, different donor or acceptor groups were systematically varied to evaluate the effect of electronic character on the 2PA properties. Additionally, it is well-known that the conjugation length exhibits a large enhancement of the 2PA absorptivity. Variation of the conjugation length in 1–19 was designed to further confirm these principles. In addition to linear structures, it has been reported that branched structures may exhibit cooperative effects on 2PA efficiency. The branched structures 22–24 (Fig. 2) were synthesized to probe this effect. It was also desirable to see if the cooperative effect was observable in linear oligomers, which was the motivation for preparation of 20 and 21 (Fig. 2).

2.2
Methodology for Synthesis of Fluorene Derivatives

The syntheses of the dyes are presented in Figs. 3 and 4. There are three major aspects of the synthesis of fluorene-based 2PA derivatives. The first is the introduction of donor or acceptor groups into the conjugation system. The second is the construction of the conjugation backbone and the third is the

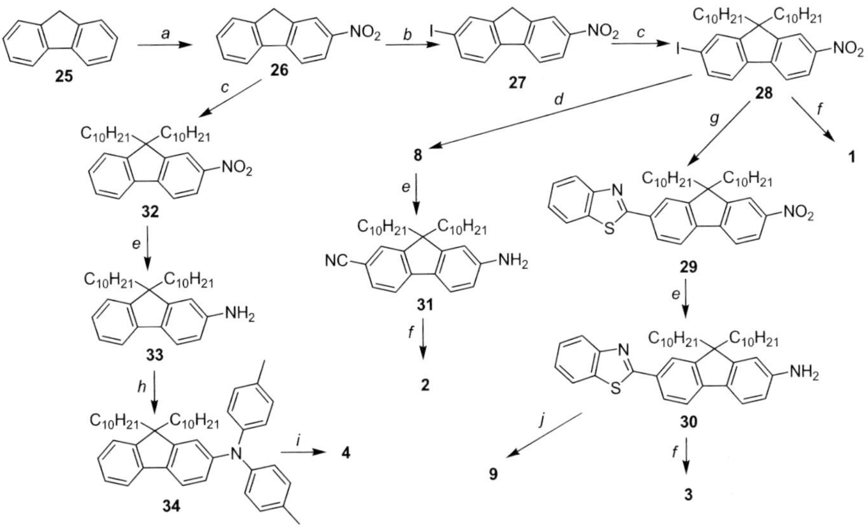

Fig. 3 Synthetic scheme for 1–4, 8 and 9. **a** HNO$_3$, HOAc, 60 °C; **b** I$_2$, HOAc, H$_2$SO$_4$, NaNO$_2$, 115 °C; **c** BrC$_{10}$H$_{21}$, KOH, KI, DMSO, 25 °C; **d** CuCN, DMF, reflux; **e** NH$_2$NH$_2$·2H$_2$O, graphite, THF/EtOH, reflux; **f** PhI or Ph$_2$NH, Cu-bronze, 18-crown-6, K$_2$CO$_3$, 1,2-dichlorobenzene, 180 °C; **g** 2-(tributylstannyl)benzothiazole, Pd(PPh$_3$)$_2$Cl$_2$, toluene, 110 °C; **h** 1-Iodo-4-methylbenzene, Cu-bronze, 18-crown-6, K$_2$CO$_3$, 1,2-dichlorobenzene, 180 °C; **i** (CH$_3$CO)$_2$O, CS$_2$, AlCl$_3$, reflux; **j** CH$_3$COCl, Et$_3$N, CH$_2$Cl$_2$, r.t.

attachment of the solubilizing groups. The first two aspects normally involve chemistry at the 2, 4, and 7 positions of fluorene while the last functionalization involves the 9 position (Fig. 3). The introduction of the donor or acceptor groups has two possibilities, i.e. directly attaching the donor or acceptor groups to the fluorenyl ring or using known intermediates with donor or acceptor groups, especially phenyl-based motifs. In the latter case, the fluorenyl group may function as the conjugation bridge.

The choice of donor and acceptor groups for our work were diphenylamino as donor and benzothiazole as acceptor, after systematic comparison of the 2PA results of compounds with a variety of donor and acceptor groups, as presented in Fig. 3. The phenyl-based diphenylamino or benzothiazole derivatives have been extensively employed in conjugated materials and their syntheses have been well documented. However, when the chemistry involves attachment directly to fluorene, it becomes much more difficult. The approach used in our group for formation of diphenylaminofluorene first involved an efficient nitration reaction of fluorene, as shown in Fig. 3, followed by reduction to a free amino group using hydrazine hydrate. The reaction is normally very efficient with a yield around 90%. The Ullmann reaction of iodobenzene, catalyzed by Cu powder in the present of a base, typically K$_2$CO$_3$ and 18-crown-6, gave satisfactory yields in most cases, e.g. the preparation of dyes 1–4 [12]. In the synthesis of dye 5, a similar Ullmann reaction was performed

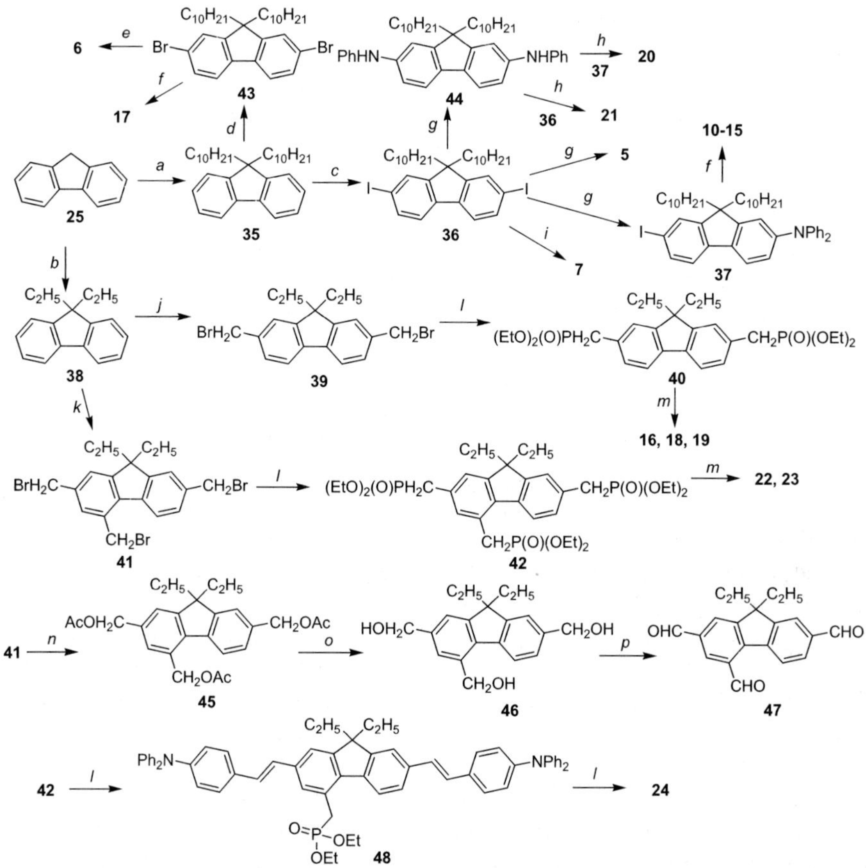

Fig. 4 Synthetic scheme for **5–7** and **10–24**. **a** BrC$_{10}$H$_{21}$, KOH, KI, DMSO, 25 °C; **b** BrC$_2$H$_5$, KOH, KI, DMSO, 25 °C; **c** I$_2$, HOAc, H$_2$SO$_4$, NaNO$_2$, 115 °C; **d** NBS, benzoyl peroxide, CCl$_4$, reflux; **e** 2-(tributylstannyl)benzothiazole, Pd(PPh$_3$)$_2$Cl$_2$, toluene, 110 °C; **f** vinyl intermediates (2-(4-vinylphenyl)benzo[d]thiazole, diethyl 4-vinylphenylphosphonate, 1-nitro-4-vinylbenzene, 2-(9,9-didecyl-7-vinyl-9H-fluoren-2-yl)benzothiazole, 9,9-didecyl-2-nitro-7-vinyl-9H-fluorene or 2-(4-(2-(9,9-didecyl-7-vinyl-9H-fluoren-2-yl)vinyl)phenyl) benzothiazole), Pd(OAc)$_2$, P(o-tolyl)$_3$, DMF, Et$_3$N, 75 °C; **g** aniline or diphenylamine, Cu-bronze, 18-crown-6, K$_2$CO$_3$, 1,2-dichlorobenzene, 180 °C; **h** **36** or **37**, Cu-bronze, 18-crown-6, K$_2$CO$_3$, 1,2-dichlorobenzene, 180 °C; **i** CuCN, DMF, reflux; **j** paraformaldehyde (2.2 equiv.), 33% HBr in HOAc, 70 °C, 20 h; **k** paraformaldehyde (10 equiv.), 33% HBr in HOAc, 70 °C, 22 h; **l** (EtO)$_3$P, reflux; **m** 4-(diphenylamino)benzaldehyde or 4-nitrobenzaldehyde, NaH, DMF, r.t.; **n** NaOAc, Ac$_2$O, HOAc, 90 °C; **o** NaOH, EtOH, 40 °C; **p** pyridinium chlorochromate, CH$_2$Cl$_2$, r.t.

successfully using diiodofluorene and diphenylamine instead. Similar methodology has been applied in the synthesis of **20** and **21**. The introduction of benzothiazole was achieved by Pd-catalyzed Stille coupling of iodofluorene derivatives with tributylstanousbenzothiazole in refluxing toluene with either tetrakis(triphenylphosphine) palladium or dichlorobis(triphenylphosphine)-palladium. The coupling afforded higher yields when an electronic withdrawing group was present at the 7-position relative to those with donor groups. All benzothiazole groups in dyes **3**, **6**, and **9** were introduced by this method.

Normally, the dialkyl groups at the 9-position of the fluorenyl ring were introduced in the early stage of the synthesis to increase the solubility of the compounds, hence facilitating the purification process. A typical process for functionalizing the 9-position with the solubilizing group is demonstrated in Fig. 3 by dialkylation of 7-iodo-2-nitrofluorene, which can be easily accomplished by generation of the fluorenyl anion with KOH in DMSO and subsequent dialkylation with 1-bromoalkane in the presence of KI at room temperature. This reaction works well regardless of the substituents at the 2- and 7-positions. This observation is of importance since, for preparation of hydrophilic fluorene derivatives, it is helpful that functionalization of the 9-position with hydrophilic chains was conducted at a late stage of the synthesis. These hydrophilic substituents may have detrimental effects on the coupling reactions, especially when metal catalysts are involved.

The extension of the conjugation system was accomplished by Pd-catalyzed Heck coupling or Horner-Emmons reactions. The conjugated backbones of **10–15**, and **17** were formed by Heck coupling of the bromo- or iodofluorene derivatives **43** and **36** with a phenyl or fluorenyl vinyl intermediate. The vinyl compounds were normally converted from corresponding benzaldehyde using classical condensation methods or using Stille coupling from iodofluorene and vinylstannous chloride. Another efficient method to form the phenylene-vinyl bridge is by Horner-Emmons coupling. The phosphates were converted from bromomethyl fluorene, which in turn was directly bromomethylated from a dialkylfluorene, such as **39** in Fig. 4. Interestingly, besides the common dibromomethylfluorene that was widely reported in literature, a trisbromomethylated compound **41** was also identified and further prepared with good yield after optimizing reaction conditions. By reacting the corresponding phosphonates with aldehydes in DMF and a base such as NaH, not only linear dyes **16**, **17**, and **19** but also branched dyes **22–24** were obtained.

3
Photophysical and Photochemical Properties of Fluorene Derivatives

3.1
Linear Absorption and Steady-state Fluorescence

Linear spectral properties of symmetrical and unsymmetrical fluorenes with different electronic structures were investigated. Electron-donating or electron-withdrawal character of the end substituents of the fluorene molecules and the symmetry of the electron density distribution play an important role in their optical properties. These structures can be represented as D-π-D, A-π-A and D-π-A molecules, where D, A, and π refer to electron donor, electron acceptor, and π-electron bridges, respectively. Absorption and fluorescence spectra for symmetrical D-π-D (**5**), A-π-A (**6**) and unsymmetrical D-π-A (**3, 11**) fluorenes in the solvents of different polarity are shown in Fig. 5. The main linear spectral characteristics of these compounds (absorption and fluorescence maxima, full width at half maximum (FWHM) of their bands, Stoke's shifts and quantum yields) are summarized in Table 1 as a function of solvent polarity. It is known that solvent polarity or orientation polarizability, Δf, can be expressed as [13]:

$$\Delta f = \frac{\varepsilon - 1}{2\varepsilon + 1} - \frac{n^2 - 1}{n^2 + 1} , \tag{1}$$

where ε and n are the dielectric constant and refraction indices of the solvent, respectively. As follows from Table 1, absorption spectra are nearly solvent invariant: the differences between λ_{max}^{abs} did not exceed 5–8 nm and were primarily determined by their molecular structure. The changes in the electronic properties of the end substituents resulted in spectral shifts of λ_{max}^{abs}. For example, the replacement of one diphenylamino-group in **5** by styrylphenylphosphoric acid diethyl ester (**11**) resulted in a red-shift of λ_{max}^{abs} by 12–16 nm.

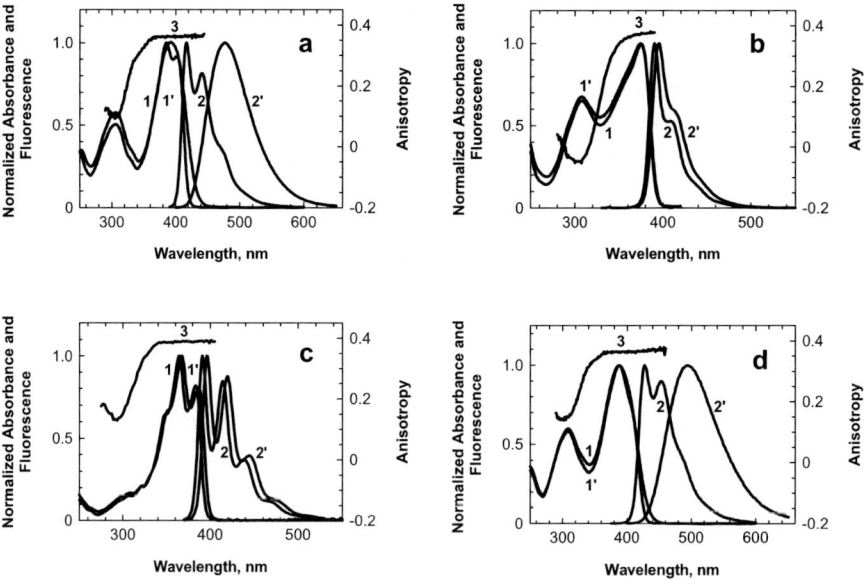

Fig. 5 Absorption (1,1'), steady-state fluorescence (2,2') and excitation anisotropy (3) spectra of **3** (**a**), **5** (**b**), **6** (**c**) and **11** (**d**) in hexane (1,2), THF (1',2') and polyTHF (3)

Table 1 Linear spectral parameters of **3**, **5**, **6** and **11**: absorption, λ_{max}^{abs}, and fluorescence, λ_{max}^{fl}, maxima, (FWHM), $\Delta\lambda^{abs}$, $\Delta\lambda^{fl}$, Stoke's shifts, fluorescence quantum yields, Φ_{FL}, and orientation polarizabilty of the solvents, Δf

Solvent	Hexane				THF				ACN			
Compound	3	5	6	11	3	5	6	11	3	5	6	11
λ_{max}^{abs}, nm	385	375	364	387	391	374	367	387	388	367	–	383
λ_{max}^{fl}, nm	417	390	391	427	478	395	396	494	532	397	–	553
$\Delta\lambda^{abs}$, cm^{-1}	3429	4421	3561	4051	2544	–	3667	3759	4130	–	–	4381
$\Delta\lambda^{fl}$, cm^{-1}	2520	1829	–	3016	3264	2030	–	3652	3826	2310	–	4359
Stoke's shift, nm	32	15	27	40	87	21	29	107	143	30	–	170
$\Phi_{FL}\cdot10^2$	70±6	40±5	95±8	60±5	55±5	50±5	96±8	90±8	90±8	70±6	–	90±8
Δf	0.0012				0.207				0.257			

The fluorescence spectra of the investigated fluorene derivatives were independent of the excitation wavelength λ_{exc} over the entire absorption region, indicative of the homogeneous spectral distribution of the emitting chromophores in the liquid solution. In contrast to absorption, the fluorescence spectra may exhibit a considerable dependence on the solvent polarity (Fig. 5b,d). This well-known solvent effect on fluorescence emission is associated with the dipolar rearrangement of the solvent shell in the excited state of the molecule [13, 14]. This spectral behavior is typical for unsymmetrical molecules that undergo a large change in dipole moment upon electronic excitation. For nonpolar solvents such as hexane, the fluorescence spectra of 3, 5, 6, and 11 exhibited well-defined vibronic structure and small Stoke's shifts. In polar solvents such as ACN, CH_2Cl_2, THF, and polyTHF, only a single broad band was observed for unsymmetrical molecules (3 and 11) with Stoke's shifts up to 170 nm (see Table 1). The values of the Stoke's shifts for compound 3, 5, 6, and 11 exhibited nearly linear dependence on Δf in accordance with the Lippert equation [13], although it was difficult to expect linear dependencies due to the changes in the shape of fluorescence spectra with corresponding changes in solvent polarity. These changes are presented in Fig. 6 for unsymmetrical fluorenes 9 and 22 [15].

Gradual changes in the shape and position of the fluorescence spectra with solvent polarity are typical for unsymmetrical fluorenes and allow consideration of the first excited singlet state of the molecule, S_1, as a quantum mixing of the Franck–Condon excited state with delocalized electronic density along the molecular axis and charge transfer state [15, 16]. Theoretically, the nature of S_1 can be characterized by the gradual change in the corresponding molecular wave function from the delocalized to the charge transfer state as a function of the solvent-controlled reaction coordinate. More details can be found in the literature [17–19].

Fluorescence quantum yields, Φ_{FL}, of the investigated fluorene derivatives (see Table 1) are relatively high ($\Phi_{FL} \approx 0.4$–1.0), and typically increase with solvent polarity (an exception is compound 3 in THF) [20, 21]. The replacement of one of the diphenylamino end groups in the symmetric molecule 5 by benzothiazole (3) or styryl phenylphosphoric acid diethyl ester (11) led to an increase in the fluorescence quantum yield in all investigated solvents. The extended symmetrical compound 16 and branched 22 also exhibited high values of $\Phi_{FL} \approx 0.9$–1.0 in all solvents, i.e. lengthening of π-conjugation may result in an increase in the fluorescence quantum yield, independent of solvent properties. The concentration dependencies of Φ_{FL} for the fluorene compounds were investigated [20] and essentially exhibited a decrease in Φ_{FL} (up to 2–3 times) at high concentration $\sim 5 \times 10^{-3}$ M in CH_2Cl_2. This behavior may be explained by the formation of nonfluorescent aggregates (e.g. dimers), however no direct evidence of aggregation of the fluorene derivatives at high concentrations was found. The absorption spectra and fluorescence lifetimes were concentration independent [20].

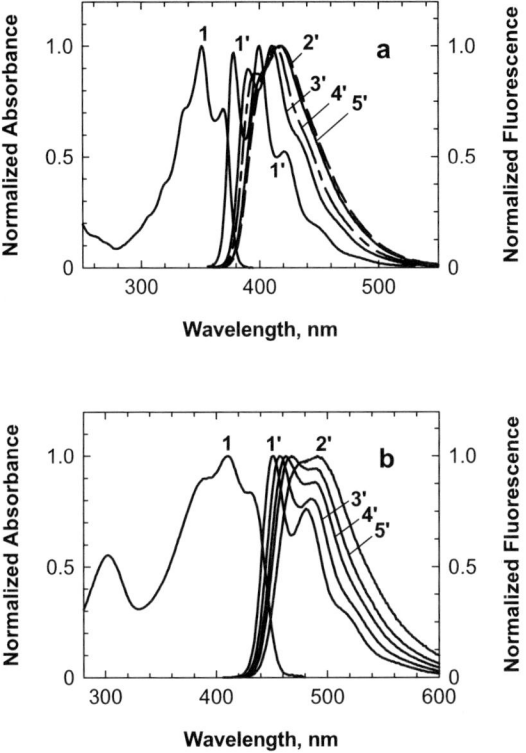

Fig. 6 Absorption (1) and fluorescence (1′–5′) spectra of **9** (**a**) and **22** (**b**) in hexane (1,1′), THF and hexane-THF mixtures (in volume percent): 80–20; 60–40; 40–60

Additional information about the nature of the absorption bands of fluorenes can be revealed from their excitation anisotropy spectra [17, 21–23]. The values of fluorescence anisotropy are commonly determined as [13]:

$$r = \frac{I_{||} - I_{\perp}}{I_{||} + 2I_{\perp}}, \tag{2}$$

where $I_{||}$ and $I_{\perp}$ are the intensities of fluorescence polarized parallel and perpendicular to the excitation light, respectively. Excitation anisotropy spectra of compounds **3**, **5**, **6**, and **11** obtained under one-photon excitation are presented in Fig. 5, curve 3. In viscous polyTHF molecular ensembles reach their limiting values [13, 17]:

$$r \approx r_{1PA} = \frac{3\cos^2\beta_{em} - 1}{5}, \tag{3}$$

where β_{em} is the angle between absorption and emission transition dipoles. This expression can be used for determination of the spectral position and orientation of several dominant electronic transitions. Constant values of

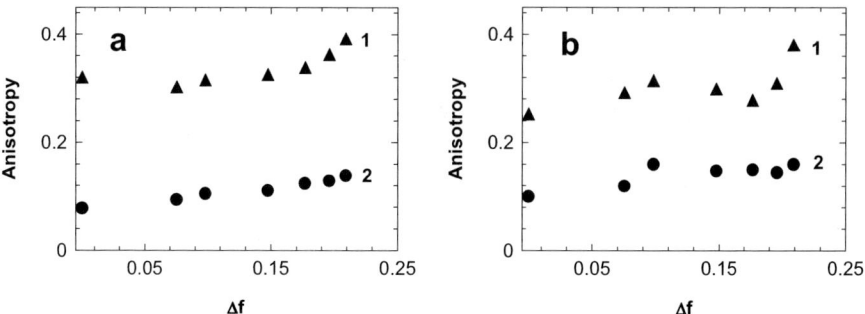

Fig. 7 Dependence of the limiting excitation anisotropy r_{1PA} on solvent polarity Δf for compounds **6** (**a**) and **9** (**b**) for excitation into the first S_1 (1), and second S_2 (2) electronic excited states ($\pm 5\%$) accuracy for each point

anisotropies in the spectral region of excitation wavelengths $\lambda_{exc} \geq 360$ nm (compounds **3**, **5**, and **11**) and $\lambda_{exc} \geq 340$ nm (**6**) correspond to the main absorption band $S_0 \rightarrow S_1$ (S_0 and S_1 are the ground and first excited states, respectively). The minimum anisotropy values of **3**, **5**, **6**, and **11** in the short wavelength region at $\lambda \approx 305$ nm (**3**, **5**, and **11**), and 295 nm (**6**), indicate the spectral position of the higher excited electronic transitions $S_0 \rightarrow S_2$. These transitions for **3**, **5**, and **11** correspond to the short wavelength absorption bands at $\lambda \approx 305$ nm, associated with the diphenylamino substituents (see Fig. 5). The limiting values of excitation anisotropy r_{1PA} exhibited a dependence on solvent polarity Δf, and noticeably increased in polar solvents at the $S_0 \rightarrow S_1$ excitation (Fig. 5, curves 1). Meanwhile, weaker dependences of r_{1PA} on Δf were observed for excitation into the second, S_2, electronic state (curves 2).

3.2
Time-resolved Emission Spectra and Fluorescence Lifetimes of Fluorene Molecules

The time-resolved emission spectra (TRES) and fluorescence lifetimes, τ_1, of the fluorene derivatives were measured in liquid solutions at room temperature with a PTI QuantaMaster spectrofluorimeter with ~ 0.1 ns temporal resolution [20]. At this resolution, all investigated fluorenes exhibited TRES which were coincident with the corresponding steady-state fluorescence spectra. As an example, TRES for compounds **3** and **11** in hexane, THF, and ACN are presented in Fig. 8 for different nanosecond delays: 0 ns (curves 2, 4, 6) and 5 ns, which modeled the steady-state condition (curves 3, 5, 7). No differences in the fluorescence spectra for these two delays were observed, indicating that all relaxation processes in the first excited state S_1 are sufficiently fast for fluorene molecules and did not exceed the time resolution of the PTI system (~ 0.1 ns).

The experimental values of τ_1, natural radiative lifetimes τ_R (calculated using the Bircks and Dyson formula [16], which is based on Strickler and

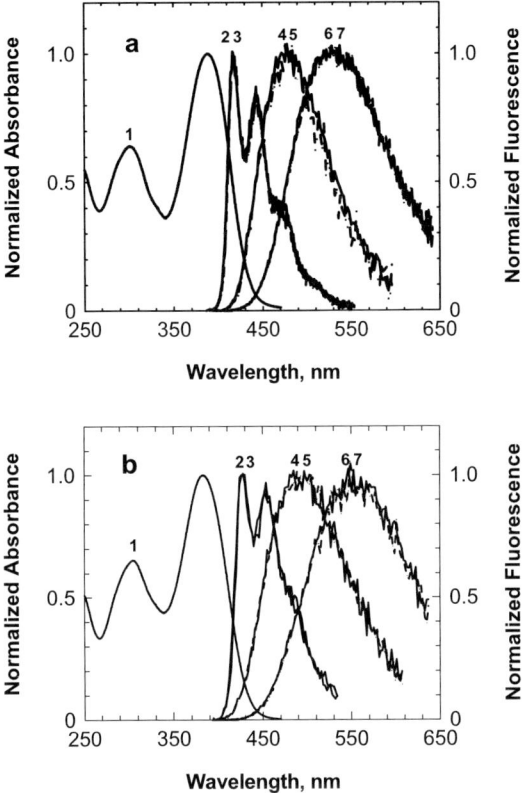

Fig. 8 Normalized absorption spectra (1) and TRES corrected on the spectral sensitivity of PMT (2–7) for compounds **3** (**a**) and **11** (**b**) for 0 ns delay (2, 4, 6) and 5 ns delay (3, 5, 7) in hexane (2, 3), THF (4, 5) and ACN (1, 6, 7)

Berg theory [24]), and calculated fluorescence lifetimes, $\tau_1^{cal} = \tau_R \cdot \Phi_{FL}$, are presented in Table 2. The observed fluorescence decays for fluorene compounds in all investigated solvents correspond to a single exponential process with typical goodness-of-fit parameters $\chi^2 \leq 1.1$ [13].The experimental values of the fluorescence lifetimes τ_1 were in reasonable agreement with calculated values τ_1^{cal} only for the nonpolar solvent (hexane). The values of τ_1 and τ_1^{cal} for polar solvents differ by 50–80% (for example, **3** in THF and **5** in ACN). This suggests that the solvent relaxation of the fluorene molecules in the first excited state (for polar solvents) may noticeably shift molecular electronic levels, so the fluorescence spectrum of this new electronic structure does not correspond to the observed absorption $S_0 \rightarrow S_1$ (even for symmetrical compound **5**). After the excitation, there is a considerable change in the excited state dipole moment (and hence, in electronic distribution) of the fluorene derivative. The solvent then reorients around this new electronic distribution, resulting in a new excited state energy level lower in energy than

Table 2 Fluorescence lifetimes τ_1, natural radiative lifetimes τ_R, and calculated fluorescence lifetimes τ_1^{cal} for compounds 3, 5, and 11 in different solvents

Solvent	Hexane			THF			ACN		
Compound	3	5	11	3	5	11	3	5	11
τ_1, ns	1.25±0.15	0.95±0.13	1.25±0.15	2.2±0.25	1.0±0.13	2.1±0.25	3.25±0.35	1.15±0.12	2.95±0.3
τ_1^{cal}, ns	1.5±0.2	0.8±0.1	1.0±0.13	1.3±0.15	1.4±0.18	2.4±0.3	2.7±0.3	1.9±0.2	2.8±0.3
τ_R, ns	2.2±0.25	2.1±0.25	1.5±0.2	2.1±0.25	2.9±0.35	2.7±0.35	3.0±0.3	2.7±0.3	3.1±0.35

the original S_1. The observed fluorescence then corresponds to this new level, not the original excited state.

In Strickler and Berg theory, it is assumed that fluorescence occurs from the original excited state without taking into account possible solvent reorientation and subsequent formation of a new lower energy excited state. Hence, the relationships between absorption and emission spectra may be more complicated than simply following Strickler and Berg theory, and τ_1 may differ from τ_1^{cal}. These results indicate the importance of considering the effect of medium on fluorescence properties for these compounds.

3.3
Lasing Potential of the Fluorene Compounds

High fluorescence quantum yields and large values of Stoke's shifts, which are known to be an important parameters for the efficient stimulated emission of organic molecules, were observed for fluorene molecules in polar solvent [17, 20, 22, 25]. Therefore, stimulated emission may be expected in these compounds and employed for practical use. The potential of lasing in the fluorene derivatives 3, 11, and 17 was comprehensively investigated [25] under pumping by the third harmonic of a picosecond Nd:YAG laser (pump wavelength, $\lambda_P = 355$ nm). An open aperture Z-scan method for nonlinear transmission measurements [26] and a picosecond pump-probe method (experimental setup is shown in Fig. 9) for excited state absorption (ESA) spectra [27] were used. In order to estimate the loss of pumping energy, the curves of nonlinear transmittance at $\lambda_P = 355$ nm were obtained for fluorene compounds 3, 11, and 17. The experimental results are shown in Fig. 10. An increase in the transmittance at high irradiance indicates that ground state absorption cross sections, $\sigma_{01}(\lambda_P)$, were larger than ESA $S_1 \rightarrow S_n$ ($n = 2, 3, ...$) cross sections, $\sigma_{1n}(\lambda_P)$, i.e. $\sigma_{01}(\lambda_P) > \sigma_{1n}(\lambda_P)$ [28]. Therefore, pump energy losses due to excited state absorption processes were small for investigated fluorenes, and the laser threshold can be achieved.

The losses in the gain region of 3, 11, and 17 were estimated from pump-probe measurements. An intensive pump pulse at $\lambda_P = 355$ nm populates the fluorescence level S_1, from which stimulated transitions in both absorption ($S_1 \rightarrow S_n$) and emission ($S_1 \rightarrow S_0$) can occur. The efficiency of these transitions was probed by a weak probe pulse tuned to the fluorescence spectral region of 3, 11, and 17 (440–630 nm). The ESA and the optical gain spectra for a probe pulse passing simultaneously with the strong pump through the solution of investigated fluorenes are displayed in Fig. 11 (curves 1–3). The same experiments were performed for Rhodamine 6G in ethanol for comparison (curve 4). From these experiments, no gain was observed for 3 and 11 in ACN over a broad spectral region, i.e. transient optical density, $\Delta D > 0$ (see curves 1, 2). The minima in these spectra at ≈ 520 nm (curve 1) and ≈ 560 nm (curve 2) are the result of two competing processes due to ESA

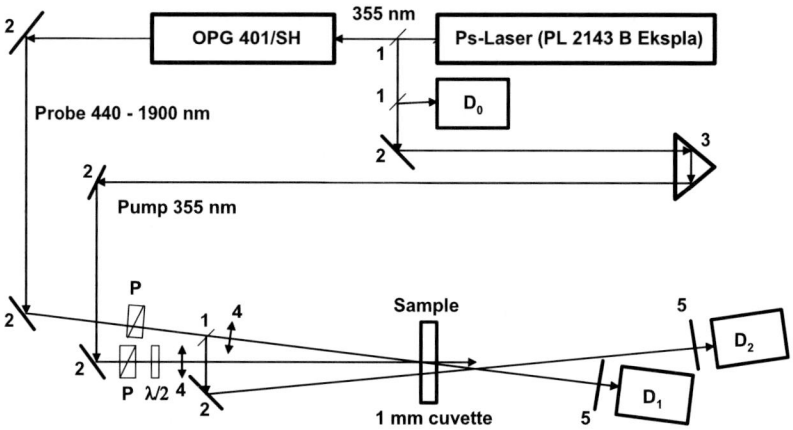

Fig. 9 Experimental setup for pump-probe measurements: 1 – beam splitters; 2 – silver mirrors; 3 – time delay line; 4 – lenses; 5 – filters; D_0, D_1, D_2 – photodetectors; P – polarizers; $\lambda/2$ – wave plate

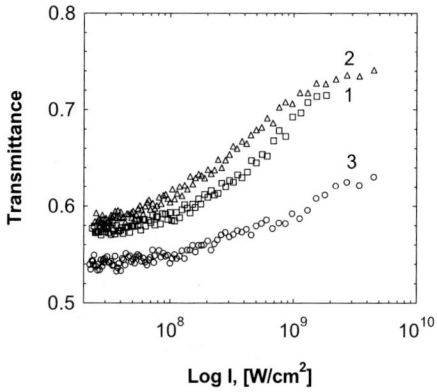

Fig. 10 Nonlinear transmittance vs. input irradiance for **3** (1), **11** (2) in ACN, and **17** (3) in THF

$S_1 \rightarrow S_n$ and stimulated emission $S_1 \rightarrow S_0$. These minima correspond to the maxima of the fluorescence bands (curves $1'$, $2'$), i.e. the maxima for potential amplification. In contrast to **3** and **11**, compound **17** in THF exhibited weak amplification ($\Delta D < 0$, curve 3). Relatively strong amplification for the well-known laser dye Rhodamine 6G in ethanol (curve 4) was observed. The linear optical density, D, of the investigated solutions in a 1 mm path length cuvette was the same for all compounds ($D \approx 0.5$ at $\lambda_P = 355$ nm). These data indicate that there is no possibility to obtain superfluorescence and lasing effects in the spectral region 440–630 nm for fluorenes **3** and **11** in ACN, due to the negative value in one-pass amplification. Small optical gain for **17** in THF (Fig. 11, curve 3) afforded lasing on the surfaces of a 1 mm cuvette (threshold irradiance $I \approx 0.5$ GW/cm^2; $D(\lambda_P) \approx 2.8$).

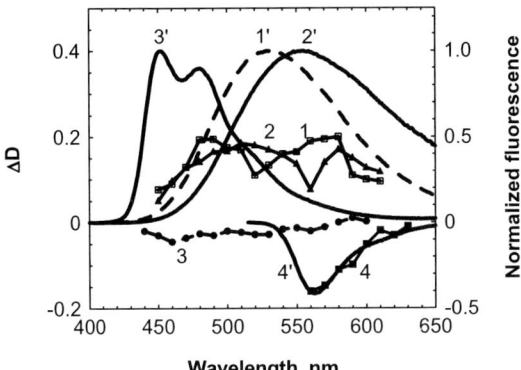

Fig. 11 Excited state absorption spectra of **3** (1) and **11** (2) in ACN, **17** (3) in THF and Rhodamine 6G (4) in ethanol. Normalized fluorescence of **3**, **11** in ACN, and **17** in THF. Fluorescence spectrum of Rhodamine 6G in ethanol is normalized to its maximum of amplification (curve 4). Pump energy $E_P \approx 350\,\mu J$ at excitation wavelength $\lambda_P = 355$ nm

The absence of optical gain for **3** and **11** in ACN can be explained by the relatively large excited state absorption of the emitted fluorescence photons from the singlet state S_1: $\sigma_{1n}(\lambda_F) > \sigma_{10}(\lambda_F)$, where $\sigma_{1n}(\lambda_F)$ and $\sigma_{10}(\lambda_F)$ are ESA and stimulated emission cross sections at the fluorescence wavelength λ_F. Note that ESA spectra strongly overlap the entire fluorescence region of **3** and **11** (Fig. 11, curves $1'$, $2'$).

Other possible reasons for the absence of the optical gain in compounds **3** and **11** were also analyzed. One of them is the population of the triplet states with the following triplet–triplet reabsorption. However, estimated intersystem crossing rates for these compounds are $\leq 3 \times 10^7\,s^{-1}$, and picosecond excitation can not lead to efficient population of the triplet states. Thus, this channel of loss is negligible. Another possible reason could be formation of photochemical products with the absorption in the fluorescence spectral region 440–630 nm of **3** and **11**. However, no absorption in this region was found experimentally by probing with a weak beam in the absence of a strong pump pulse.

The results presented above indicate that the unsymmetrical fluorene derivatives with diphenylamino substituents, in spite of large Stoke's shifts and high fluorescence quantum yields, do not exhibit superfluorescence and lasing effects. On the other hand, this result shows the potential of fluorene derivatives for their use in photon counting devices, where a linear dependence of fluorescence intensity on pump energy is needed. Figure 12 shows the dependence of the relative fluorescence intensity, I_{FL}, on the absorbed energy of the excitation pulse for **11** in ACN. The solution was excited in a 1 mm path length quartz cuvette at $\lambda_P = 355$ nm. The optical density of this solution was nearly $D(\lambda_P) \approx 4$ and corresponded to the full absorption of pump energy over the entire range of pump energies. The relative intensity of the fluo-

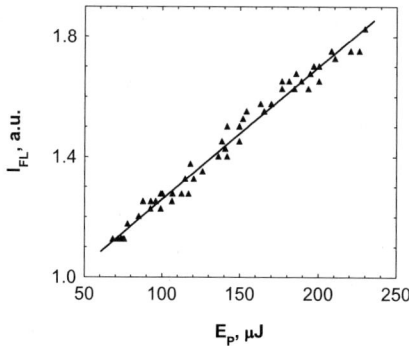

Fig. 12 Dependence of the relative fluorescence intensity I_{FL} on the absorbed pump energy E_P at $\lambda_P = 355$ nm for compound 11 in ACN

rescence I_{FL} was measured in the transverse direction relative to the pump beam [20, 29]. The linear dependence provides the ability to determine the number of photons in one excitation pulse at a high pump irradiance from 1 to 4 GW/cm². Thus, unsymmetrical fluorene derivatives can be used as photon counters for high intensity laser pulses.

3.4
Excited-state Absorption and Anisotropy Properties of Fluorene Derivatives

The electronic structure of fluorenes and the development of their linear and nonlinear optical structure-property relationships have been the subject of intense investigation [20–22, 25, 30, 31]. Important parameters that determine optical properties of the molecules are the magnitude and alignment of the electronic transition dipole moments [30, 31]. These parameters can be obtained from ESA and absorption anisotropy spectra [32, 33] using the same pump-probe laser techniques described above (see Fig. 9). A comprehensive theoretical analysis of a two beam (pump and probe) laser experiment was performed [34], where a general case of induced saturated absorption anisotropy was considered. From this work, measurement of the absorption anisotropy of molecules in an isotropic ensemble facilitates the determination of the angle between the $S_0 \to S_1$ (pump) and $S_1 \to S_n$ (probe) transitions. The excited state absorption anisotropy, r_{abs}, is expressed as [13]:

$$r_{abs} = \frac{k_{||} - k_\perp}{k_{||} + 2k_\perp} , \tag{4}$$

where $k_{||}$ and $k_\perp$ are the excited state absorption coefficients of the probe pulse for parallel and perpendicular orientations with respect to the pump beam polarization. The ESA spectra were obtained as a transient optical density, $\Delta D_{||,\perp} = (k_{||,\perp} \cdot L)/\ln 10$, where $L = 1$ mm (the path length of the cuvettes in all pump-probe measurements). The excited state absorption co-efficients of the probe pulse, $k(\theta)$, for an arbitrary angle, θ, between the pump and probe polarization can be derived in the case of unsaturated absorption as [22, 34]:

$$k(\theta) = k_0[\cos^2 \theta \cdot (3 \cos^2 \beta_{abs} - 1) - \cos^2 \beta_{abs} + 2],\qquad(5)$$

where k_0 is a constant value and β_{abs} is the angle between $S_0 \rightarrow S_1$ and $S_1 \rightarrow S_n$ transition dipoles. As follows from Eq. 5, the function $k(\theta)$ remains symmetrical for the whole range of angles $0° \leq \beta_{abs} \leq 90°$, and direct angular measurements of the shape of $k(\theta)$ do not give any information on β_{abs}. Taking into account that $k(0°) = k_{||}$ and $k(90°) = k_{\perp}$, substitution of Eq. 5 into Eq. 4 gives an expression identical to Eq. 3, well-known for fluorescence anisotropy:

$$r_{abs} = \frac{3 \cos^2 \beta_{abs} - 1}{5}.\qquad(6)$$

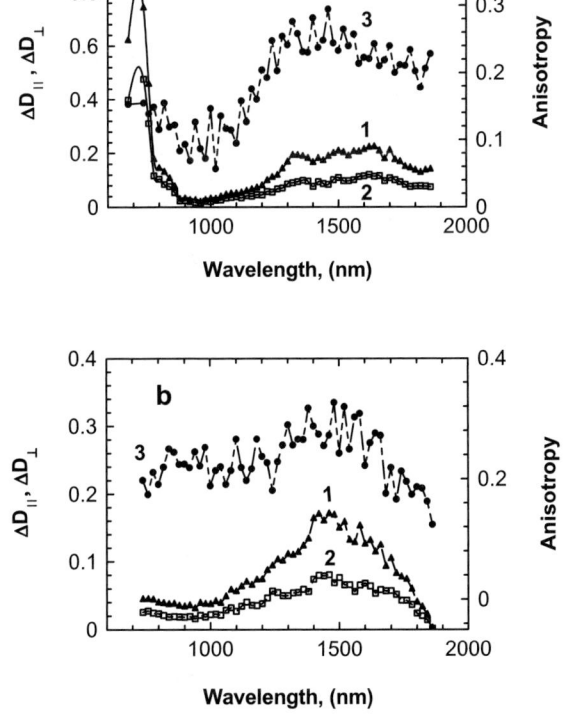

Fig. 13 Excited-state absorption ($\Delta D_{||}$ (1), $\Delta D_{\perp}$ (2)) and absorption anisotropy (3) spectra of compounds **6** (**a**) and **9** (**b**) in polyTHF

The values of β_{abs} were determined from the experimentally measured absorption anisotropy with Eq. 6. In viscous polyTHF, the rotational movement of dye molecules on a $\approx 100\,\mathrm{ps}$ time scale is assumed to be negligible, and thus, does not reduce the limiting value of anisotropy. For the concentrated fluorene solutions ($5 \times 10^{-4}\,\mathrm{M}$, 1 mm cuvette), the anisotropy r_{abs} was not affected by Förster depolarization mechanisms [13] due to the short time delay, τ_{D}, between probe and pump pulses when $\tau_{\mathrm{D}}/\tau_1 \ll 1$ [35].

The ESA and absorption anisotropy spectra of fluorenes 6, 9, 16, 22 are presented in Figs. 13–14. According to this data, compounds 6 and 9 exhibited broad ESA bands in the spectral region from 1300–1700 nm. These bands can be attributed to one or several electronic transitions. The ESA anisotropy spectra of 6 and 9 revealed a possible range of angles between $S_0 \rightarrow S_1$ and $S_1 \rightarrow S_n$ electronic transitions in this spectral region of $25° \leq \beta_{\mathrm{abs}} \leq 35°$. In addition, fluorene 6 exhibited a strong ESA band at $\lambda \approx 720\,\mathrm{nm}$ with a relatively small value of absorption anisotropy $r_{\mathrm{abs}} \approx 0.15$ corresponding to $\beta_{\mathrm{abs}} \approx 40°$.

The ESA spectra of 16 and 22 (Fig. 14, curves 1, 2) exhibited strong bands at $\lambda \approx 900\,\mathrm{nm}$ with $r_{\mathrm{abs}} \leq 0.1$. These absorption bands were assigned to elec-

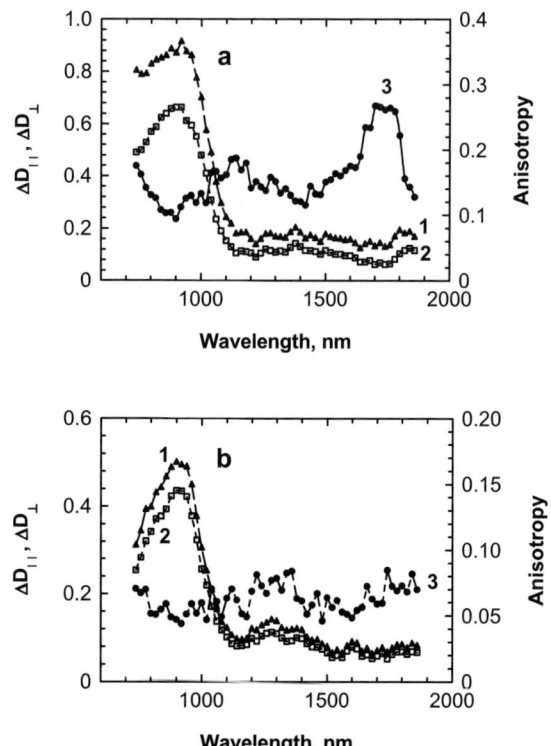

Fig. 14 Excited state absorption ($\Delta D_{\parallel}$ (1), $\Delta D_{\perp}$ (2)) and absorption anisotropy (3) spectra of compounds 16 (**a**) and 22 (**b**) in polyTHF

tronic transitions involving the diphenylamino substituents. Another weak ESA band of 7 was observed at $\lambda \approx 1750$ nm (with $r_{abs} \approx 0.26$) and extracted from the excitation anisotropy spectrum. The angles between the $S_0 \rightarrow S_1$ and $S_1 \rightarrow S_n$ electronic transitions of 7 in polyTHF: $35° \leq \beta_{abs} \leq 50°$ indicated a preference for a nonparallel orientation. The excitation of 22 at $\lambda = 355$ nm primarily corresponded to the second electronic transition $S_0 \rightarrow S_2$ revealed in the steady-state excitation anisotropy spectrum ($r_{abs} \approx 0.08$ at 330 nm $\leq \lambda \leq 370$ nm) [22]. The orientation of this transition was $\beta_{abs} \approx 45°$ relative to the ESA dipole $S_1 \rightarrow S_n$. In this case the alignment of the $S_0 \rightarrow S_1$ and $S_1 \rightarrow S_n$ transition dipole moments in 22 can not be determined from these experimental data. Thus, ESA and absorption anisotropy spectra revealed some features of the electronic molecular structures which can be used for optimization of the investigated fluorene derivatives for nonlinear optical applications.

3.5
Two-photon Absorption Properties of Fluorene Derivatives

3.5.1
General Aspects of Two-photon Absorption

The phenomenon of two-photon absorption (2PA) can be presented as a process of simultaneous absorption of two photons under high intensity irradiation, resulting in one excited molecule [1]. The investigations of 2PA previously undertaken are of great interest for a wide variety of emerging applications [3, 36–42]. The nature of 2PA can be described based on the interaction of molecular electrons with an optical field. On a microscopic level the displacement of molecular electronic charge under the electric field, E, is related to the induced molecular dipole moment μ:

$$\mu = \alpha \cdot E + \frac{1}{2} \beta \cdot E \cdot E + \frac{1}{6} \gamma \cdot E \cdot E \cdot E + \dots, \tag{7}$$

where α is a linear (first order) molecular polarizability, β and γ are the second and third order polarizabilities, respectively. It can be shown that only the imaginary parts of odd-order terms of polarizabilities ($\alpha, \gamma, \dots$) in Eq. 7 contribute to the dissipative processes, such as one-photon (α) or two-photon (γ) absorption [43, 44]. This means that the lowest-order nonlinear absorption (i.e. 2PA) will be described by the imaginary part of γ. In the case of isotropically oriented molecules and parallel polarization of photons with equal frequencies, ω (degenerate 2PA), the value of γ is related to the 2PA absorption cross section, δ_{2PA}, as [44–46]:

$$\delta_{2PA} = \frac{3L^4 \hbar \omega^2}{2n^2 c^2 \varepsilon_0} \mathrm{Im} \langle \gamma(-\omega; \omega, -\omega, \omega) \rangle, \tag{8}$$

where n and c are the refractive index of the medium and the speed of light in a vacuum, respectively, ε_0 is the permittivity of free space and L is the local field factor. The latter describes the interaction between neighboring molecules (brackets indicate an average of γ over isotropic medium) [47].

In order to establish the relationship between the molecular structure and its 2PA parameters, the tensor components of the third order polarizability, γ_{ijkl}, need to be determined. Several quantum-chemical approaches are employed to calculate tensor components γ_{ijkl} and corresponding cross sections δ_{2PA} [48–50]. One of the most widely used methods is the Sum-Over-State (SOS) approach developed by Orr and Ward [51], which involves the calculation of the molecular wavefunctions and the values of permanent and transition dipole moments for all electronic states which contribute to the polarizability. These contributions from the excited states can then be summed based on time-dependent perturbation theory, and the tensor components γ_{ijkl} can be determined. In the general case the expression for γ_{ijkl} is relatively complicated [51] due to summation over all of the excited states of the molecule. Therefore, for practical use a simplified model expression for γ_{ijkl} can be obtained from the full SOS expression. According to this simplified model, the ground state of the molecule $|g\rangle$ is strongly coupled to a single one-photon allowed state $|e\rangle$, and there are also several two-photon allowed states $|e'\rangle$ that are strongly coupled to $|e\rangle$. In the case of degenerate 2PA, when considering only resonant terms and $xxxx$-component of γ_{ijkl} the full SOS expression can be reduced to the positive dipolar, two-photon and negative terms [8]:

$$
\gamma_{xxxx}(-\omega;\omega,-\omega,\omega) = 2 \cdot \left\{ \begin{array}{ll}
\dfrac{\mu_{ge}^x \cdot \bar{\mu}_{ee}^x \cdot \bar{\mu}_{ee}^x \cdot \mu_{eg}^x}{(E_{eg} - \hbar\omega)^2 (E_{eg} - 2\hbar\omega)} & D \\[3mm]
+ \dfrac{\mu_{ge}^x \cdot \bar{\mu}_{ee}^x \cdot \bar{\mu}_{ee}^x \cdot \mu_{eg}^x}{(E_{eg}^* - \hbar\omega)(E_{eg} - 2\hbar\omega)(E_{eg} - \hbar\omega)} & D \\[3mm]
- \dfrac{\mu_{ge}^x \cdot \mu_{eg}^x \cdot \mu_{ge}^x \cdot \mu_{eg}^x}{(E_{eg} - \hbar\omega)^3} & N \\[3mm]
- \dfrac{\mu_{ge}^x \cdot \mu_{eg}^x \cdot \mu_{ge}^x \cdot \mu_{eg}^x}{(E_{eg}^* - \hbar\omega)(E_{eg} - \hbar\omega)^2} & N \\[3mm]
+ \displaystyle\sum_{e'} \dfrac{\mu_{ge}^x \cdot \bar{\mu}_{ee'}^x \cdot \bar{\mu}_{e'e}^x \cdot \mu_{eg}^x}{(E_{eg} - \hbar\omega)^2 (E_{e'g} - 2\hbar\omega)} & T \\[3mm]
+ \displaystyle\sum_{e'} \dfrac{\mu_{ge}^x \cdot \bar{\mu}_{ee'}^x \cdot \bar{\mu}_{e'e}^x \cdot \mu_{eg}^x}{(E_{eg}^* - \hbar\omega)(E_{e'g} - 2\hbar\omega)(E_{eg} - \hbar\omega)} & T,
\end{array} \right\} \quad (9)
$$

where μ_{ij}^x $(i,j = g, e)$ are the x-components of the transition dipole moments between corresponding electronic states, $\bar{\mu}_{ij}^x = \mu_{ij}^x - \delta_{ij}\mu_{gg}^x$ $(i,j = e, e'$, δ_{ij} is a Kronecker symbol) are the changes in the permanent dipole moments relative to the ground state, and $E_{ig} = E_j - E_g - i\Gamma_{jg}$ ($j = e, e'$, and E_i, E_g and Γ_{jg} are the energy of corresponding electronic states and damping factors, respectively) [52]. Only D and T terms contain two-photon resonances with $|e\rangle$ and $|e'\rangle$ states and, therefore, can contribute to 2PA cross section. As follows from Eqs. 7 and 8, in the case of resonance two-photon excitation into the $|e\rangle$ state ($\hbar\omega \approx \frac{E_e - E_g}{2}$ and $\Gamma_{eg} \ll \frac{E_e - E_g}{4}$):

$$\delta_{2PA} \sim \frac{\mu_{ge}^2 \cdot \Delta\mu_{ge}^2}{\Gamma_{eg}}, \qquad (10)$$

where $\Delta\mu_{ge} = \mu_{ee} - \mu_{gg}$ is the change of the permanent dipole moment between the ground state $|g\rangle$ and the excited state $|e\rangle$. It is obvious from the equation that for centrosymmetric molecules which are characterized by zero values of $\Delta\mu_{ge}$, two-photon excitation into the $|e\rangle$ state is strictly forbidden. In contrast, unsymmetric compounds with strong one-photon absorption and large values of $\Delta\mu_{ge}$ can exhibit efficient 2PA under excitation into the first excited $|e\rangle$ state [53].

For the case of resonance, two-photon excitation into the two-photon allowed states $|e'\rangle$ ($\hbar\omega \approx \frac{E_{e'} - E_g}{2}$ and $\Gamma_{eg} \ll \frac{E_e - E_g - (E_{e'} - E_g)/2}{2}$), the value of 2PA cross section can be expressed as:

$$\delta_{2PA} \sim \frac{\mu_{ge}^2 \cdot \mu_{ee'}^2}{\Gamma_{e'g}(E_e - E_g - (E_{e'} - E_g)/2)^2}. \qquad (11)$$

As follows from the equation, the efficiency of 2PA depends on the values of the corresponding one-photon transition dipoles and the detuning energy $E_e - E_g - (E_{e'} - E_g)/2$.

The above description of 2PA processes corresponds to a simplified essential state model which is widely used to analyze the structure-properties relationship of 2PA materials [8]. In practice, this theoretical approach is in a good agreement with experimental data for different types of organic compounds [54–56].

3.5.2
Two-photon Absorption Efficiency of Fluorenes

Comprehensive experimental investigations of 2PA processes in fluorene derivatives were performed by Hales et al. [53, 56–59] with open aperture Z-scan [26], two-photon induced fluorescence [60] and femtosecond white-light continuum pump-probe methods [61]. For degenerate two-photon excitation, the experimental 2PA spectra of symmetrical and asymmetrical fluorenes are presented in Figs. 15 and 16. These spectra were obtained with the combination of open aperture Z-scan and two-photon fluorescence methods [57]. For centrosymmetric molecules, two-photon transitions from

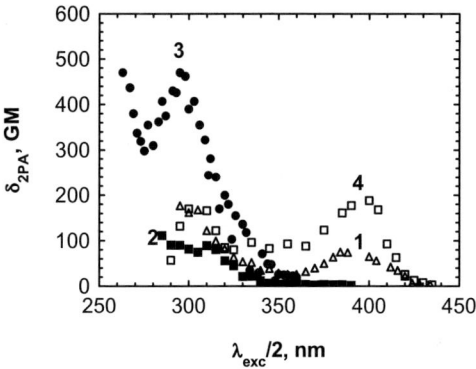

Fig. 15 Degenerate 2PA spectra of compounds **3** (1), **5** (2), **6** (3), and **11** (4) in hexane

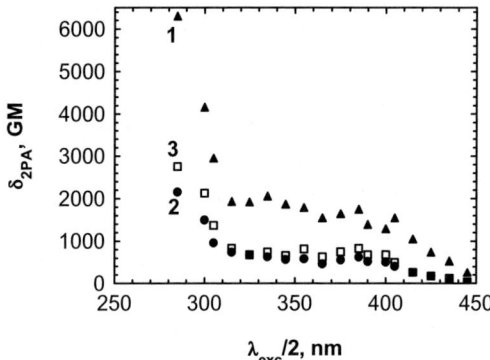

Fig. 16 Degenerate 2PA spectra of compounds **17** in CH_2Cl_2 (1), **16** (2) and **22** (3) in cyclohexane

the ground (A_g) to the first excited electronic state (as a rule, $1B_u$) are strictly forbidden in accordance with the dipole selection rules [62]. Therefore, symmetrical fluorene derivatives **5** and **6**, which are close to a centrosymmetric structural type, exhibited relatively weak 2PA (10–30 GM) in the spectral range of the main one-photon allowed absorption band $S_0 \rightarrow S_1$ (see Figs. 5, 6 and 15, curves 1, 4). Under two-photon excitation into the second electronic state $S_0 \rightarrow S_2$ (two-photon allowed), the efficiency of 2PA increased up to 100 GM for **5** and 450 GM for **6**. The spectral position of the second electronic state can be estimated from the one-photon excitation anisotropy spectra (Fig. 5a,c, curves 3). Unsymmetrical fluorenes **3** and **11** possess comparable values of 2PA (100–200 GM) in the broad spectral range corresponding to both $S_0 \rightarrow S_1$ and $S_0 \rightarrow S_2$ two-photon excitations (Fig. 15, curves 2, 3). Long symmetrical **16**, **17**, and branched **22** fluorene derivatives exhibited a complicated 2PA spectrum without well defined absorption maxima in the measured spectral range, which overlaps at least two electronic transitions.

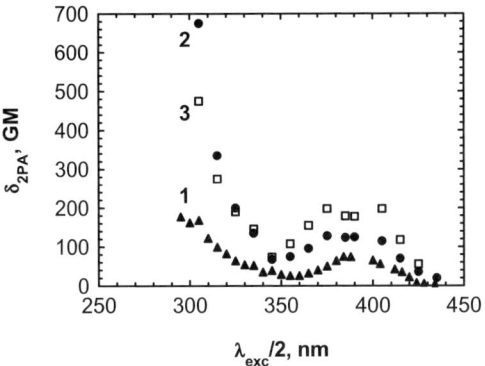

Fig. 17 Degenerate 2PA spectra of compound 3 in hexane (1), CH$_2$Cl$_2$ (2) and ACN (3)

For the investigated fluorene compounds, the values of 2PA cross sections δ_{2PA} increased with the lengthening of the conjugated system, which is typical for many 2PA chromophores [53, 55, 63]. The comparison of δ_{2PA} for **6** and **17** (with the same benzothiazole end substituents and different lengths of the conjugated chains) revealed a nearly 10-fold increase in δ_{2PA} for the whole range of excitation wavelengths. One of the possible explanations of this effect concerns the number of π electrons in chromophore system participating in 2PA process [64]. An increase in the number of π electrons leads to an increase in δ_{2PA}. Electron affinities of the terminal groups in fluorenes structure also determine the efficiency of 2PA. As can be seen in Fig. 15 (curves 1, 3), electron- withdrawing (benzothiazole) end substituents in fluorene **6** resulted in 3–4 times larger δ_{2PA} in comparison to corresponding values for compound **5** with electron- donating (diphenylamino) end groups. This suggests that under the excitation, center-to-periphery charge transfer in symmetrical compound **6** corresponds to the larger transition dipole moment and higher efficiency of 2PA relative to the corresponding parameters of **5** with periphery-to-center charge transfer. Finally, the polarity of the solvent can also effect δ_{2PA}. 2PA spectra of **3** in the solvents of different polarity (Δf) are shown in Fig. 17 [57]. As follows from these data, the efficiency of 2PA can increase with solvent polarity 2 - 3 times. A more pronounced effect was observed for unsymmetrical molecules. The most probable explanation of this behavior concerns the possibility of larger intramolecular charge transfers produced upon excitation of fluorenes in more polar solvents. This can lead to the enhanced values of dipole moments $\mu_{ee'}$ of the higher electron transitions and larger changes in the permanent dipoles $\Delta\mu_{ge}$ [57]. According to equations from Sect. 3.5.1, the value of δ_{2PA} is quadratically dependent on $\Delta\mu_{ge}$ and $\mu_{ee'}$, therefore, a noticeable increase in 2PA efficiency can be observed in the polar solvents.

3.5.3
Fluorescence Anisotropy of Fluorenes under Two-photon Excitation

Anisotropy properties of the molecular fluorescence under two-photon excitation reflect the nature of 2PA processes and may provide additional information on the electronic structure of the molecules, including the peculiarities of the 2PA mechanism. In general, the measurements of two-photon fluorescence anisotropy are more sensitive than at one-photon excitation due to a broader range of anisotropy values [13] that in some cases provide extra advantages for practical applications of 2PA [65].

As was shown above, the 2PA spectra of symmetrical and unsymmetrical fluorene derivatives exhibit a complex nature of 2PA bands in the spectral region 280–420 nm (see Figs. 15, 16). In general, the nature of 2PA processes should be reflected in anisotropy spectra. These spectra, $r_{2PA}(\lambda_{exc})$, can be obtained with Eq. 1 in the same way as $r_{1PA}(\lambda_{exc})$, however, with two-fold longer excitation wavelengths (560–840 nm).

The excitation anisotropy spectra $r_{2PA}(\lambda_{exc})$ obtained for symmetrical (**6, 17**) and unsymmetrical (**9, 11**) fluorenes under two-photon excitation are presented in Fig. 18 (curves 1, 1′). As follows from these experimental data, the anisotropy values are relatively high and remain nearly constant over the entire spectral region (covering at least two electronic transitions): $r_{2PA} \approx 0.53$ (**6**), $r_{2PA} \approx 0.51$ (**9**), $0.38 \leq r_{2PA} \leq 0.47$ (**11**) and $0.40 \leq r_{2PA} \leq 0.55$ (**17**). Observed values of $r_{2PA}(\lambda_{exc})$ differ from the one-photon anisotropy values $r_{1PA}(\lambda_{exc})$ which exhibited considerable changes at excitation into the different electronic states (see Fig. 5). A possible explanation of this unusually invariant behavior of $r_{2PA}(\lambda_{exc})$ can be provided based on the simplest 2PA mechanism with only one intermediate $|e\rangle$ and final $|e'\rangle$ electronic state [65].

For this three-state model the values of δ_{2PA} can be determined by the Eqs. 10 and 11. In this simple model the chromophore system of the molecule can be modeled by two arbitrarily oriented linear oscillators, μ_{ge} and $\Delta\mu_{ge}$ (for excitation into the first excited electronic state S_1), or by the μ_{ge} and $\mu_{ee'}$ (for excitation into the final electronic state S_f), which simultaneously absorb two photons and transfer their energy to the emission oscillator, μ_{eg}^{fl}. It has been shown that the limiting value of fluorescence anisotropy r_{2PA} can be written as [23]:

$$r_{2PA} = \frac{18\cos(\xi/2 - \beta')\cdot\cos(\xi/2 + \beta')\cdot\cos(\xi) - 7\cos^2(\xi) + 1}{7[2\cos^2(\xi) + 1]}, \tag{12}$$

where ξ are the angles between the absorption dipoles μ_{ge}, $\mu_{ee'}$ (or μ_{ge}, $\Delta\mu_{ge}$), and β' is the angle between the emission dipole μ_{eg}^{fl} and the bisecting line of the angle ξ. In the case of parallel orientation of the absorption oscillators $\xi = 0$, and β' becomes equal to the angle between absorption and

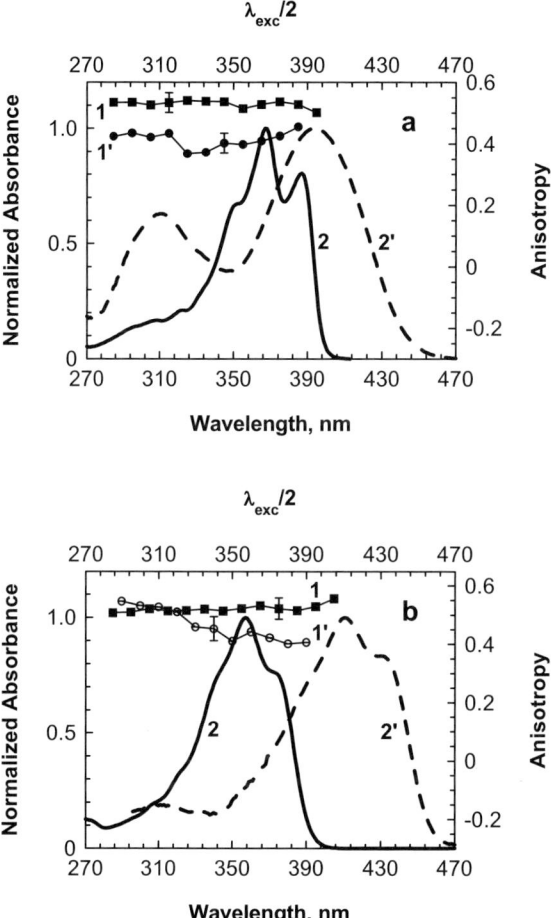

Fig. 18 Two-photon excitation anisotropy spectra (1, 1′, *top scales*) for compound **6** (**a**, **1**), **11** (**a**, 1′), **9** (**b**, **1**) in polyTHF and **17** (**b**, 1′) in silicon oil. *Solid lines* (2, 2′) are the corresponding linear absorption spectra

emission dipoles β_{em}. Then Eq. 9 reduces to the well-known expression [13]:

$$r_{2PA} = \frac{6\cos^2\beta_{em} - 2}{7}. \tag{13}$$

As can be seen from the Eq. 12, the two-photon anisotropy magnitude range of $-0.29 \leq r_{2PA} \leq 0.57$ at $0° \leq \beta_{em} \leq 90°$ is broader as compared to the corresponding one-photon anisotropy range of $-0.2 \leq r_{1PA} \leq 0.4$ (see Eq. 3).

The values of two-photon anisotropy as a function of ξ, calculated by Eq. 9 at several angles $\beta_{em} = \beta' + \xi/2$, are shown in Fig. 19. The parameters β_{em} were chosen in accordance with the experimentally observed range of the corresponding angles for fluorene derivatives in polyTHF [15, 21, 22, 25]. As follows from Fig. 19, a broad range of ξ between $0°$ and $\sim 40°$ corresponds to the relatively small changes in r_{2PA} from 0.45 to 0.6 (i.e. relative changes $\leq 25\%$). This means good agreement between the calculated anisotropy values of three-state model and the experimental data for symmetrical and unsymmetrical fluorenes at the excitation into the second (final) electronic state.

Two-photon excitation into the main absorption band, however, is expected to reveal a different behavior for symmetrical and unsymmetrical molecules. It is known that for symmetrical fluorenes μ_{01} is oriented perpendicular to $\Delta\mu$ (i.e. $\xi = 90°$), which should lead to the low values of $r_{2PA} = 0.14$ (see Eq. 9 and Fig. 19), in contradiction to the experimental results. For unsymmetrical fluorenes, the calculated angles between $\Delta\mu$ and μ_{01} typically show a large deviation from the perpendicular orientation ($30° \leq \xi \leq 40°$), and theoretical values of r_{2PA} are in a good agreement with the experimental data. Thus, the three-state model approximation can be used for the description of two-photon anisotropy of asymmetrical fluorenes under $S_0 \rightarrow S_1$ excitation, but it is inadequate to model the symmetrical ones. High values of r_{2PA} (0.4–0.55) for symmetrical molecules for excitation into the main absorption band can be explained by taking into account the effect of vibronic coupling, which can partly break the symmetry of the molecule, resulting in the noticeable changes of the angle ξ between absorption oscillators. The deviation of ξ from $90°$ can be sufficiently large (up to $\xi = 40°$) [23] due to large differences between the values of μ_{01} and $\Delta\mu$ ($\mu_{01} \sim 10$ D and $\Delta\mu \leq 1$ D for chromophores close to centrosymmetric) [66, 67]. In this case, a small change in the orientation of μ_{01} may cause a relatively large deviation in the orientation of $\Delta\mu$. Thus, this deviation from a perpendicular orientation may explain the disagreement between experimental (0.4–0.55) and calculated (0.14) values of r_{2PA} for symmetrical fluorenes at $S_0 \rightarrow S_1$ excitation. Comprehensive quantum-chemical calculations and further experimental investigations are needed for a deeper understanding of the observed invariant two-photon anisotropy behavior.

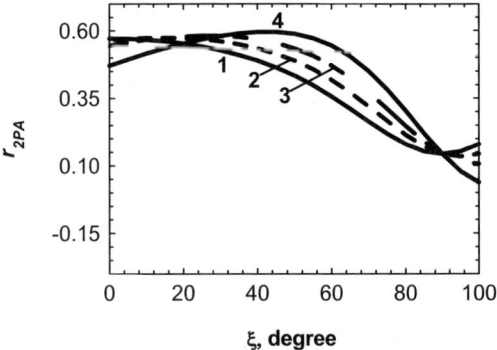

Fig. 19 Dependences $r_{2PA} = f(\xi)$ calculated for the three-state model at: $\beta_{em} = 0°$ (1); $5°$ (2); $10°$ (3); $20°$ (4)

3.6
Photochemical Properties of Fluorene Derivatives

Photochemical stability of organic molecules at high intensity irradiation is one of the important issues in the development of linear and nonlinear-optical technologies. Therefore, the main photochemical parameters of new optical materials need to be determined for their possible applications. Photodecomposition of the molecules can occur as a result of one-photon, two-photon, or two-step successive absorption within a single pulse of laser excitation. Under these conditions, organic molecules can be excited into high electronic states and undergo different photochemical reactions, such as photoionization and bond fission [68]. In general, these photoprocesses may differ from the reactions of the molecule in its lowest excited state produced by low intensity irradiation. The comparison of photobleaching processes under one- and two-photon excitation may help to reveal their natures and determine the peculiarities of their photochemical behaviors. In general, relatively little is known about the photochemical reactivity of fluorenes, even though a number of fluorene derivatives have potential use in nonlinear optical technologies [53, 58, 59, 63]. Thus, the investigation of the photostability of new 2PA fluorene derivatives is important to understand their possible applications and limitations.

3.6.1
Photochemical Properties of Fluorenes Under One-photon Excitation

The photochemical stability of the molecules is characterized by the quantum yield of photodecomposition, $\Phi = N/Q$ [69], where N and Q are the numbers of decomposed molecules and absorbed photons, respectively. The photochemical properties of the fluorene derivatives were investigated in different organic solvents (hexane, CH_2Cl_2, ACN, and polyTHF) at room temperature by the absorption and fluorescence methods and comprehensively described [70–72]. These methods are based on measurements of the temporal changes in the steady-state absorption and fluorescence spectra during irradiation. For the absorption method, the quantum yield of the photodecomposition under one-photon excitation, Φ_{1PA}, can be obtained by the equation [73]:

$$\Phi_{1PA} = \frac{N_A \cdot [D(\lambda, 0) - D(\lambda, t_0)]}{10^3 \varepsilon(\lambda) \cdot \int\limits_{\lambda} \int\limits_{0}^{t_0} I_0(\lambda)[1 - 10^{-D(\lambda,t)}] \, d\lambda \, dt}, \tag{14}$$

where N_A is Avogadro's number, $D(\lambda, 0)$, $D(\lambda, t_0)$, $\varepsilon(\lambda)$, t and λ are the initial and final optical densities of the solution, the extinction coefficient (in $M^{-1} cm^{-1}$), the irradiation time, and the excitation wavelength, respectively,

t_0 is the total time of irradiation, and $I_0(\lambda)$ is the irradiance of excitation. In the case of the fluorescence method, Φ_{1PA} can be determined as [64]:

$$\Phi_{1PA} = \frac{1 - F(t_0)/F}{\int\limits_{\lambda} \int\limits_{0}^{t_0} I_0(\lambda)\sigma(\lambda)[F/F]\,d\lambda\,dt}, \tag{15}$$

where F and $F(t_0)$ are the initial and final fluorescence intensities (in relative arbitrary units) and $\sigma(\lambda)$ is the one-photon absorption cross section (in cm^2). It should be noted that some additional requirements for the experimental conditions need to be fitted for these measurements (for details see [70–72]).

The experimental data on the photodecomposition quantum yields of the fluorene derivatives under one-photon excitation into the main absorption band are presented in Table 3. The values of Φ_{1PA} were obtained for the range of molecular concentrations 10^{-3} M $\leq$ C $\leq 10^{-6}$ M, as well as for air-saturated and deoxygenated solutions [70, 71, 73, 74]. As follows from Table 3, the values of Φ_{1PA} strongly depend on the solvent properties. In polar aprotic ACN, the values of Φ_{1PA} exhibit dependence on fluorene concentration and the oxygen content in the solvent, indicative of second order photoreactions with an important role of the singlet and triplet molecular oxygen [71]. The highest photochemical stability in ACN was observed for compound 3 in the deoxygenated solution, with $\Phi_{1PA}^d \approx 3\cdot10^{-6}$. The quantum yields of photobleaching in nonpolar hexane exhibited an unexpected opposite dependence on the oxygen content in the solutions, i.e. a 2–3 times increase in deoxygenated solutions. This is suggestive of efficient quenching of potentially reactive excited state species, such as a triplet state of the fluorene molecule [70]. The lowest photostability was observed for compound 5 in CH$_2$Cl$_2$, with $\Phi_{1PA} \approx (1.0-1.5)\cdot10^{-2}$. The values of the photochemical quantum yields in CH$_2$Cl$_2$ were independent of fluorene and oxygen concentrations, consistent with first order photoreactions. Thus, in this case, the main mechanism can be associated with an electron transfer process from the amine group leading to formation of stable cation radicals [68]. This mechanism is nearly independent of oxygen concentration, resulting in many different photoproducts with broad absorption in the visible and near-IR range. Long linear fluorene derivative 16 and branched 22 unexpectedly exhibited the highest level of photostability, with $\Phi_{1PA} \approx 2\cdot10^{-6}$ in polyTHF, which increased more than 10-fold in deoxygenated solutions, revealing an important role of molecular oxygen in the photoreactions.

3.6.2
Photochemical Properties of Fluorenes Under Two-photon Excitation

The photochemical stability of the fluorene derivatives under two-photon excitation into the main absorption band was investigated by the fluorescence method described above, with the use of a Clark-MXR, CPA2010 femtosecond laser, an optical parametric generator/amplifier (TOPAS 4/800, Light Conversion) for the excitation and a PTI spectrofluorimeter for the detection [73]. The quantum yields of photodecomposition were calculated from the experimental data in a Gaussian spatial and temporal beam profile approximation [70, 75] and are presented in Table 4. The analysis of these data revealed a weak dependence of Φ_{2PA} on fluorene concentration in hexane and CH_2Cl_2, very similar to the photochemical behavior of **3** and **5** upon one-photon excitation (see Table 3). Thus, a dominant role of the first order photoreactions can also be assumed. The values of photodecomposition quantum yields

Table 3 Quantum yields of the fluorenes photodecomposition under one-photon excitation into the main absorption band ($\lambda_{exc} = 360$ nm and 420 nm*), in air saturated (Φ_{1PA}) and deoxygenated (Φ_{1PA}^d) solutions with different concentrations C

Compound	Solvent	Concentration $C \cdot 10^5$ M	$\Phi_{1PA} \cdot 10^5$	$\Phi_{1PA}^d \cdot 10^5$
3	Hexane	0.3	3.5 ± 1	14 ± 4
		2	5.0 ± 1.5	12 ± 4
		8	4.0 ± 1.3	15 ± 5
		30	4.5 ± 1.5	12 ± 4
		130	5.0 ± 2.0	–
	CH_2Cl_2	0.3	2.5 ± 0.8	8.0 ± 2.5
		2	3.0 ± 1.3	10 ± 3
		8	4.0 ± 1.2	13 ± 4
		30	3.5 ± 1	12 ± 4
		130	4.0 ± 1.5	–
	ACN	2.4	35 ± 8	0.27 ± 0.06
		0.16	3 ± 1	–
5	Hexane	0.3	35 ± 8	50 ± 12
		2	34 ± 8	60 ± 15
		8	40 ± 10	150 ± 50
		30	46 ± 12	100 ± 25
		130	34 ± 8	–
	CH_2Cl_2	0.3	1500 ± 500	1500 ± 500
		2	1200 ± 400	1200 ± 400
		8	1000 ± 300	1000 ± 300
		30	1000 ± 300	1300 ± 400
		130	1000 ± 300	–
	ACN	4.5	65 ± 15	18 ± 5
		0.24	6.5 ± 1.5	–
11	ACN	2.5	39 ± 10	1.6 ± 0.4
		0.16	23 ± 6	–
16	polyTHF	1.8	0.20 ± 0.05*	0.012 ± 0.003*
22	polyTHF	1.9	0.21 ± 0.05*	0.02 ± 0.05*

Table 4 Quantum yields Φ_{2PA} of photodecomposition under two-photon excitation into the main absorption band ($\lambda_{exc} = 720$ nm and 840 nm*) in air saturated solutions with different concentrations C

Compound	Solvent	Concentration $C \cdot 10^5$ M	$\Phi_{2PA} \cdot 10^5$
3	Hexane	0.3	3.0 ± 1.0
		2	3.5 ± 1.2
		8	3.0 ± 1.0
		30	4.5 ± 2.0
		130	5.0 ± 2.2
	CH$_2$Cl$_2$	0.3	90 ± 45
		2	80 ± 40
		8	200 ± 100
		30	150 ± 50
		130	200 ± 100
5	Hexane	0.3	23 ± 10
		2	20 ± 10
		8	22 ± 10
		30	20 ± 10
		130	30 ± 15
	CH$_2$Cl$_2$	0.3	700 ± 300
		2	1000 ± 500
		8	1700 ± 800
		30	800 ± 400
		130	1400 ± 700
7	polyTHF	1.8	0.19 ± 0.07*
8	polyTHF	1.9	0.15 ± 0.06*

of fluorenes under one- and two-photon excitation were sufficiently close, indicating similar mechanisms of the photochemical reactions. Only one exception was observed for compound **3** in CH$_2$Cl$_2$, when the values of Φ_{2PA} were 30–50 times larger than those obtained for low intensity one-photon irradiation. This photochemical behavior can be explained by an additional one-photon reabsorption process from the first excited state followed by efficient photoreaction from a higher electronically excited state [70].

In conclusion, it should be noted that fluorene derivatives **16** and **22**, with large two-photon absorption cross sections, high fluorescence quantum yields and high photochemical stability under one- and two-photon excitation are outstanding candidates for various linear and nonlinear optical applications, especially 3D fluorescence bioimaging.

4
Applications

4.1
Two-photon Fluorescence Imaging

Two-photon fluorescence microscopy (2PFM), first demonstrated by Denk et al. [3] in 1990, is based on the condition of a fluorophore to simultaneously absorb two lower energy photons in a single quantum event to induce an electronic excitation that is normally accomplished by a single higher energy photon. 2PFM is inherently characterized by the high spatial localization of this excitation event via the quadratic relationship or nonlinear dependence of two-photon absorption and fluorescence on the intensity of the incident light. Practically, this means fluorescence occurs only at the focal volume, and as near-IR laser irradiation is used as the excitation source, deeper imaging into optically thick tissue, with spatially restricted photobleaching and phototoxicity of the imaging specimen are achieved. In a medium containing one-photon absorbing chromophores, significant absorption occurs all along the path of a focused beam of suitable wavelength light. This can lead to out-of focus excitation. In a two-photon process, negligible absorption occurs except in the immediate vicinity of the focal volume of a light beam of appropriate energy. This allows spatial resolution about the beam axis as well as radially, which circumvents out-of-focus absorption and is the principle reason for two-photon fluorescence imaging [3]. The tenets of 2PA enable investigations of complex biological problems and experiments on living samples not possible with other imaging techniques.

Two-photon and higher multiphoton microscopies, utilizing judicious choices of optical probes, have yielded sophisticated and unparalleled imaging methods and techniques relative to what is achievable from linear fluorescence imaging methods. In the case of laser-scanning two-photon excited fluorescence microscopy, selecting the optimal fluorophore is critical, and data of the 2PA and emission spectra for commonly used fluorophores and bioindicators have been generated to facilitate the selection process [60, 76]. Particular molecules can undergo upconverted fluorescence through nonresonant two-photon absorption using near-IR radiation, resulting in an energy emission greater than that of the individual photons involved (upconversion). The use of a longer wavelength excitation source for fluorescence emission affords advantages not feasible using conventional UV or visible fluorescence techniques, e.g. deeper penetration of the excitation beam and reduction of photobleaching. Argon ion (488 nm) and frequency-doubled Nd:YAG (532 nm) lasers are the commonly used light sources for conventional (single-photon) laser scanning confocal microscopy due to their ready availability and low cost. Such light sources require fluorophores with strong absorbance near these wavelengths.

Two-photon laser scanning fluorescence microscopy systems, on the other hand, are generally configured with a Ti:sapphire laser with 80–120 fs pulse output in the near-IR region (700–900 nm). A reasonable absorption maximum for such chromophores is 380–420 nm (facilitating the use of near-IR femtosecond sources in the range 760–840 nm), since the 2PA λ_{max} will be approximately twice the wavelength of the single-photon λ_{max}. Commercial fluorophores are far from being optimized for use in two-photon fluorescence microscopy. Many of these compounds are conventional UV-excitable fluorophores, exhibiting low 2PA cross sections δ, of the order of ~ 10 GM units with a few exhibiting ~ 100 GM units in this wavelength area. Only recently has research been reported on the design and development of very efficient 2PA dyes possessing between 100 and several thousand GM units [7, 55, 77]. A more salient feature of a fluorophore for its use in 2PFM is its two-photon excited fluorescence action cross section $\eta\delta$, the product of the 2PA cross section δ and the fluorescence quantum yield η. Usually expressed in GM units, the action cross sections of these dyes are typically lower than the 2PA cross section.

Higher values of $\eta\delta$ enable detection of lower dye concentrations with reduced laser power required for imaging, resulting in improved signal collection due to suppressed autofluorescence and less phototoxic effects on the sample. Hence, synthetic efforts focused on preparing new nonlinear optical probes specifically engineered to exhibit higher δ are expected to outperform standard fluorophores currently in use for 2PFM imaging. For these applications, more criteria need to be met. They must be photostable and able to withstand high irradiances under one-photon excitation (1PE) and two-photon excitation (2PE). They must be nontoxic, having minimal cytotoxic effects, especially on living samples, and they must be hydrophilic, or aqueous compatible. Currently, efforts are increasingly directed toward developing and evaluating analogous compounds with greater compatibility or relevance to biological environments, such as increasing hydrophilicity as well as specificity.

The well-characterized 2PA fluorene dye **3** has been proven to be a very efficient 2PA and fluorescence compound. Recently it has also been successfully used as a 2PF probe for biological imaging [78]. Two-photon excited fluorescence microscopy images of **3** staining fixed rat cardiomyoblasts is demonstrated in Fig. 20. The fluorescent image of the fluorophore-stained cells in Fig. 20D reveals higher contrast and greater signal under the same excitation and power exposure as that of the control cells without fluorophore, which only showed modest autofluorescence, as shown in Fig. 20C. Two-photon induced fluorescence was observed predominantly from the cytoplasmic region, consistent with the images collected with epi-fluorescence in Fig. 20B. These results lend credence to our motivation toward developing fluorene-based reagents for multiphoton bioimaging applications.

Probe **3** is a highly hydrophobic compound that only achieved limited solubility in a DMSO/water mixture. To prepare a more hydrophilic version of dye

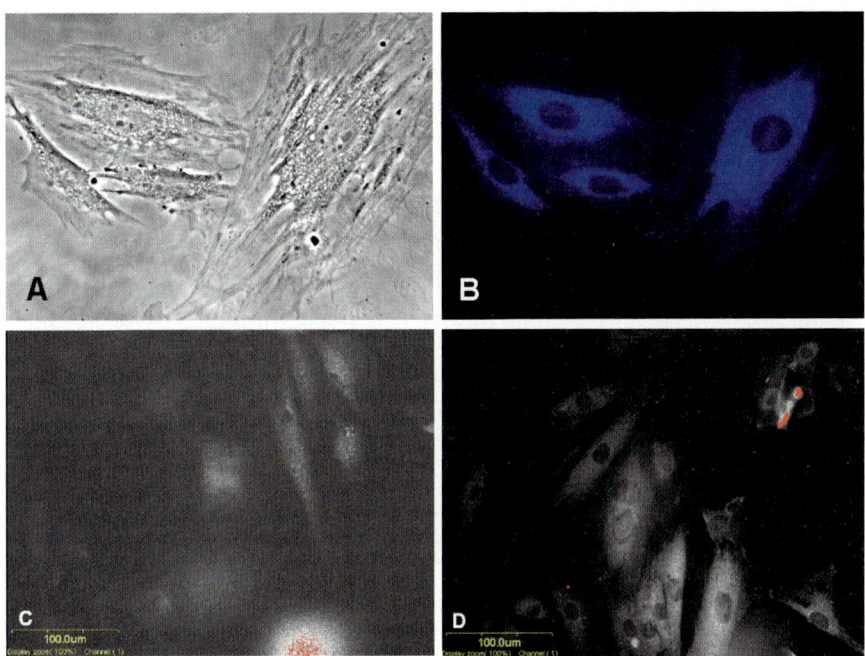

Fig. 20 **A** Bright field transmission, **B** epi-fluorescent microscopy images of H9c2 cells stained with 2PA dye **3**. **C** 2PFM images of a blank containing no fluorophore and exhibiting some autofluorescence and **D** cells stained with fluorene **3** upon 800 nm fs excitation. *Red spots* demark signal saturation. Bright field transmission and epi-fluorescence microscope images were collected on a Nikon Eclipse E600 upright microscope. Two-photon excited fluorescence microscopy images were obtained on a modified Olympus IX 70 inverted microscope and Fluoview laser scanning unit accommodating a 10-W Verdi pumping a Ti:sapphire crystal of a Mira 900 [79]

3, a similar conjugated system was functionalized by ethylene-oxy chains to form **49** (Fig. 21), which possesses improved solubility in polar solvents, especially DMSO/water mixtures. This dye was successfully used to stain NT2 cells, a human cancer cell line. Fluorescent images of NT2 cells incubated with fluorophore **49** were collected through a modified CFP filter set (DAPI exciter Ex377/50 with CFP 458DM, Em483/32) to accommodate the spectral profile of this dye using an Olympus X81 DSU microscopy system. The fluorescence appeared predominately in the cytoplasmic region as punctuated structures, possibly exhibiting preferential staining for cytoplasmic organelles such as mitochondria. This was further confirmed by observation of the fluorescence overlapping in the live NT2 cells co-stained with **49** and a mitochondria fluorescent probe MitoTracker Red CMXRos [80]. The preferential staining for organelles of dye **49** may extend its utility in fluorescence imaging of animal cells. A two-photon induced fluorescence image of the fixed NT2 cells, incubated with compound **2** upon exposure to 800 nm excitation (160 fs, 10 mW,

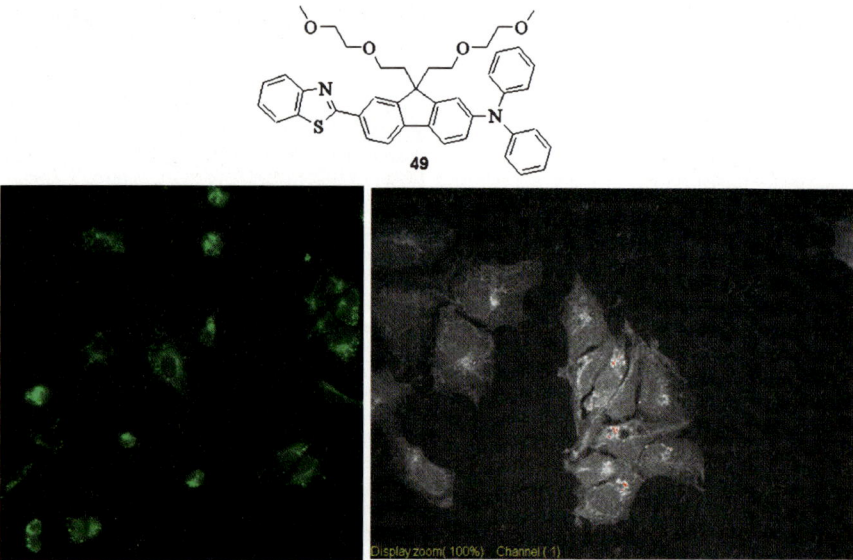

Fig. 21 Structures of hydrophilic 2PA fluorophores **49** used for cell staining, and confocal microscopy images of live NT2 cells incubated with a two-photon absorbing hydrophilic probe **49** and two-photon induced fluorescence image of fixed NT2 cells (60× oil) stained with the hydrophilic probe **49**

76 MHz, 60× oil objective) using a modified Olympus IX70 microscope, is shown in Fig. 21. Red spots in the image indicate signal saturation. The observed fluorescence appeared predominantly in the cytoplasmic region. This reiterates the utility of this compound as an efficient 2PA fluorescent contrast agent for cellular imaging.

Currently, amine reactive fluorescent probes specifically used to covalently label biomolecules that exhibit high 2PA cross sections and sufficient action cross sections are rare and appear to be limited to dipyrrylmetheneboron difluoride dyes [81, 82]. Therefore, the necessity to incorporate efficient reactive 2PA fluorophores with high action cross sections for covalent attachment to biomolecules within the fluorophore design strategy is timely and fulfills an appropriate requirement that coincides with increasing usage of two-photon excitation fluorescence imaging methods and techniques in the life sciences. Efforts directed toward preparing reactive fluorescent reagents have been initiated with the synthesis of an amine-reactive tag **50** (Fig. 22). The isothiocyanate functionality reacts with aliphatic amine groups, including the N-terminus of proteins and the ε-amino groups of lysines. A model probe adduct **51** was first prepared by reacting **50** with n-butylamine to test its reactivity as an amine-reactive fluorescent label, and its spectroscopic and labeling properties have been determined. The conjugate exhibits a high fluorescence quantum yield of 0.74 in DMSO, while the fluorescence quantum

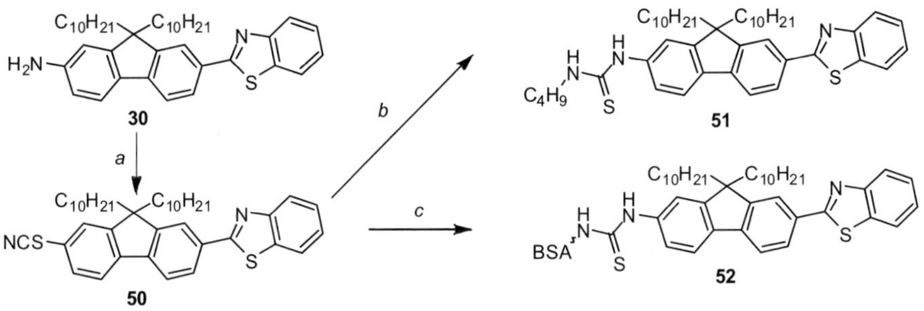

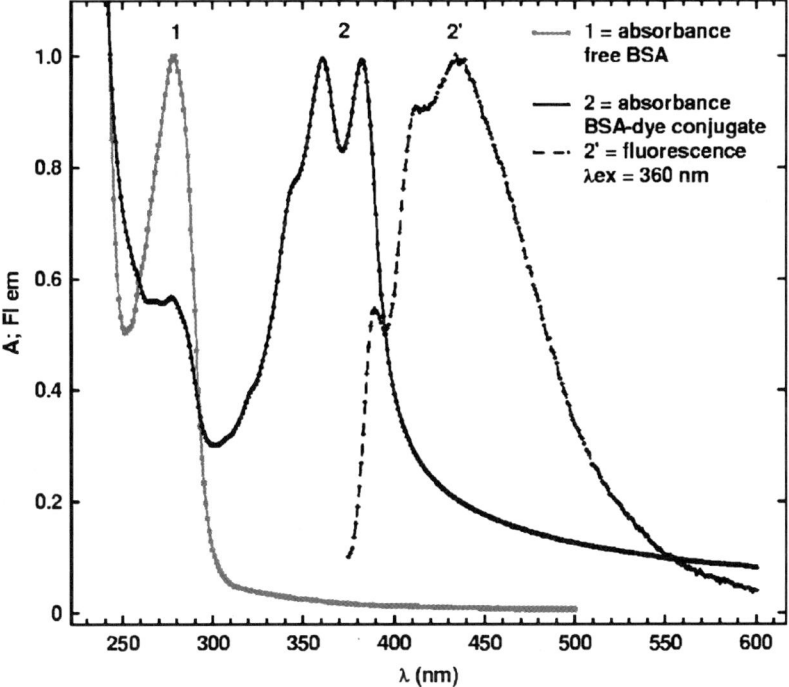

Fig. 22 Preparation of the model adduct and bioconjugate **51** and **52** with the amine-reactive fluorenyl reagent. **a** CSCl$_2$, CaCO$_3$, H$_2$O/CHCl$_3$, 0 °C; **b** n-butylamine, r.t.; **c** BSA, r.t.

yield of the reactive, unconjugated reagent **50** in the same solvent was only 0.02.

Finally, a model protein bioconjugate **52** was prepared with the reactive fluorophore **50** and bovine serum albumin (BSA). The conjugate was identified spectrophotometrically and its steady state fluorescence emission spectra

Fig. 23 Normalized absorption spectra of the free BSA protein (1), BSA-dye **50** conjugate (2) and steady state fluorescence emission spectrum of the BSA-dye conjugate (2′)

subsequently obtained. Different molar ratios of the reactive dye to protein under different reaction times were performed to assess the reactivity of the dye for its degree of labeling (DOL). The DOL is a key parameter to establish, as over-labeling of a fluorescent tag may interfere with the biological activity of a particular protein. A DOL value of 2.2 to 3.4 was obtained with **50**, a typical range for amine-reactive probes.

The normalized absorption and steady state fluorescence emission spectra of the BSA-dye conjugate in PBS buffer (pH 7.2) are shown in Fig. 23. The conjugate displays absorption peaks corresponding to that of the BSA protein in the shorter wavelength range $\lambda_{max} = 280$ nm, as well as that of the fluorescent tag in the longer absorption range $\lambda_{maxima} = 360$ and 380 nm. The fluorescence emission of the bioconjugate is broad and exhibits an appreciable Stokes shift. A bathochromic shift in the fluorescence emission was observed in the BSA-dye conjugate, relative to that of the free reactive fluorophore. Initial demonstration of the fluorenyl tag **50** as an amine-reactive 2PA fluorophore is part of our continuing program toward developing more hydrophilic fluorene derivatives with higher action cross sections.

4.2
Two-photon 3D Data Storage

Recently, with the increase of storage data density both magnetic and conventional optical data storage technologies, in which individual bits are stored as distinct magnetic or optical changes on the surface of a recording medium, are approaching physical limits beyond which individual bits may be too small or too difficult to store and retrieve. Storing information throughout the volume of a medium, not just on its 2D surface, offers an intriguing high-capacity alternative, achievable by using many layers with relatively large marks (i.e. greater than 1 micrometer). This approach can potentially provide efficient storage at densities significantly higher than those that are likely to be available from magnetic media. Two-photon 3D data storage is one of the most promising techniques to meet these demands.

The premise of the technology is that 2PA of laser light can be used to initiate photochemical processes that alter the local optical properties of a material. The 2PA is confined to the focus volume (voxel), resulting in 2PA induced modulation of optical properties that is highly localized within the focal volume. Thus, 3D spatial control is achievable with high precision. Demonstrated capabilities include the recording and reading of media with more than 100 data layers, recording tracks of $2 \times 2\ \mu m^2$ data marks, and the construction of several proof-of-principle portable readout systems. Two-photon recorded 3D optical storage technology development has been predicted to provide disk drive systems with high capacity (100–1000 GB/disk) and data transfer rates of 1–10 GB/sec, using inexpensive, easily manufactured polymer media [83]. A range of different materials and processes has been investigated for 2PA-based 3D data storage since the Rentzepis' pioneering work of 1989 [2].The first compound attempted in our group for two-photon 3D data storage is a well-documented fulgide compound **53** [84, 85], as shown in Fig. 24.

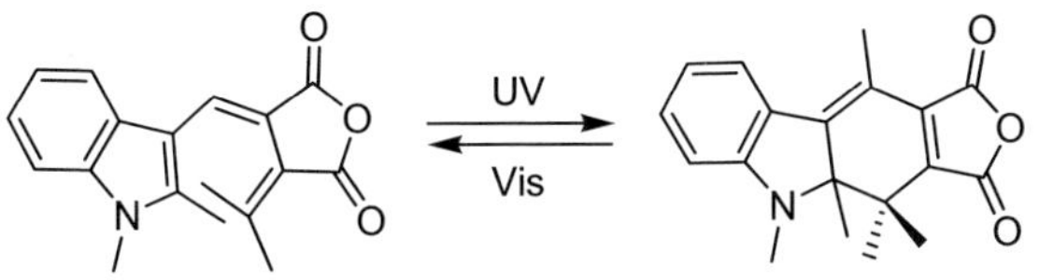

Fig. 24 Photoisomerization of indolylfulgide **53** (open form λ_{max} = 385 nm, closed form λ_{max} = 590 nm)

Fulgide-type compounds are known to undergo photoisomerization from a colorless to highly colored isomer [86]. The thermally and photochemically stable fulgide-type compounds have been reported that underwent numerous single-photon photochemical isomerization (color) and reversion cycles without significant degradation [87]. Optical data recording potentials as high as 10^8 bits cm^{-2} have been reported for fulgide-type materials. Under two-photon excitation, formation of the fulgide photoisomer **53** closed form was monitored as a function of time. Plots of absorbance at 585 nm ($\log I_0/I$) versus time were linear for the formation of the ring-closed photoisomer. The two-photon induced photoisomerization rate constants were 2.53×10^{-3}– 0.3×10^{-3} and 6.99×10^{-3}–0.5×10^{-3} s^{-1} at irradiation intensities of 3.5 and 7.0 mW, respectively. We have demonstrated 2-D interferometric recording using a Mach–Zehnder interferometry setup and a Clark CPA2001 775 nm femtosecond laser as the irradiation source (Fig. 25). Photoinduced changes were observed in the regions of high light intensity (bright interference fringes) in a thin film of poly(styrene)-fulgide **53** composite, demonstrating a proof-of-principle for effecting photochromic transformations in localized regions (Fig. 25) as a model for holographic information storage.

Recently, a new photochromic system, diarylethene, was investigated for two-photon data storage application [88] due to its excellent fatigue resistance, picosecond switching time, high photoisomerization quantum yields, and absence of thermal isomerization [89, 90]. The novel system is based on the modulation of the fluorescence emission of a highly efficient two-photon absorbing fluorescent dye and a photochromic diarylethene, providing a non-destructive readout method with $>10^4$ readout cycles. The storage medium consists of a photochromic compound (diarylethene 1), 1,2-bis(2-methylbenzo[b]thiophen-3-yl)hexafluorocyclopentene, and an efficient two-photon absorbing dye **17** (Fig. 26) [88].

Interestingly, neither the open nor the closed form of **54** displays significant fluorescence (fluorescence QY$_{OF}$ = 0.02, QY$_{CF}$<0.02). In a manner similar to that reported by Castellano [91], however, the closed form of diarylethene

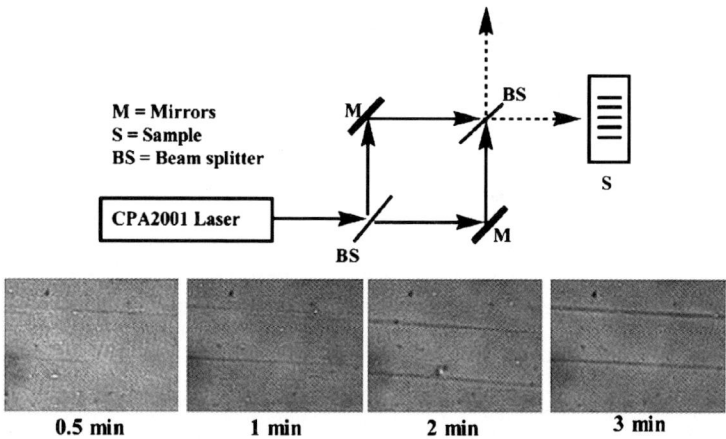

Fig. 25 Schematic of a Mach–Zehnder interferometer (*upper*) for 2-D recording via two-photon photochromism (beam splitters are 50 : 50 at 45°, laser exposure time) 0.5–3 min, output beam angle 2°. *Dark lines* in image (*lower*) result from high intensity bright fringe-induced photoisomerization of fulgide **1** in a polymeric film (13 μm line width and 155 μm line spacing)

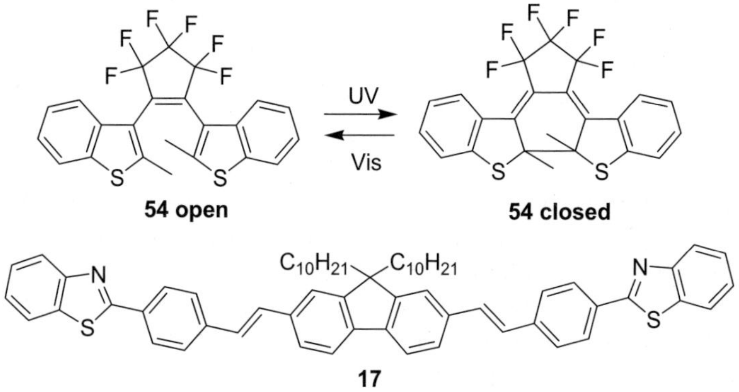

Fig. 26 Molecular structures of the open and closed forms of the diarylethene **54** and the 2PA fluorene derivative **17**

54 can be used as a photochromic energy transfer quencher of the emission of fluorescent dye **17** (Fig. 27). Compared to previously reported diarylethene fluorescent switches (in which fluorescent dyes are covalently linked to the diarylethene) [92], fluorescence modulation from non-covalently attached fluorescent dyes is a particularly intriguing approach due to the relative synthetic ease and versatility in materials selection.

Stable uniform films were obtained by mixing closed form diarylethene **54** and fluorene **17** with poly[methylmethacrylate-*co*-(diethylvinylbenzyl-phosphonate)] (PMMA-*co*-VBP) [93]. The deep red films coated on glass

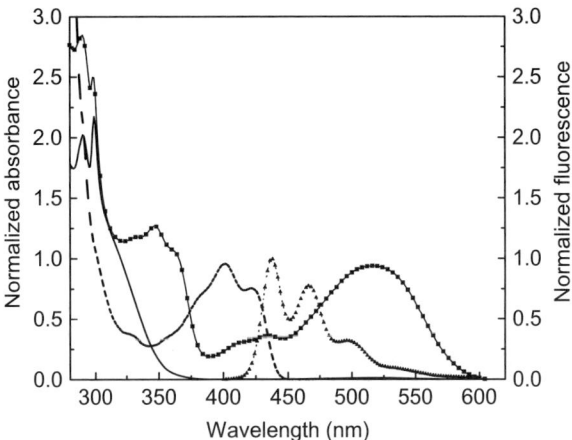

Fig. 27 Normalized absorption spectra of the open form (—) and photostationary state (■) of diarylethene **54** (irradiating at 254 nm), and absorption (- - -) and emission spectra (▲) of fluorene derivative **17** ($\lambda_{exc} = 400$ nm)

slides were approximately 40 μm. The fluorescence from fluorene **17** was effectively quenched in this film, and the fluorescence was recovered after photo ring-opening of **54**, giving the fluorescence contrast for readout. Films containing 22 wt % of diarylethene **54** and 1 wt % fluorene **17** (relative to the polymer) afforded the best contrast for a given exposure time and thickness. Shown in Fig. 28 are the single-photon and two-photon readout data recorded by single-photon writing.

Using a modified Olympus Fluoview FV300 two-photon microscope equipped with a tunable femtosecond laser (tuned to 800 nm in this experiment), two-photon readout was convincingly demonstrated (Fig. 28). In particular, reading of the memory (data) is achieved by measuring the two-photon up-converted fluorescence of fluorene **17** at 800 nm as a function of position. At this wavelength, fluorene **17** has a δ_{2PA} of 1185 GM, while the δ_{2PA} of the closed form of diarylethene **54** (~120 GM from picosecond Z-scan experiments) is one order of magnitude lower than that of fluorene **17**. Thus, when a weak 800 nm femtosecond laser (<10 mW) was utilized for the readout process, strong two-photon fluorescence from **17** was obtained. Meanwhile, this incident intensity is too weak to cause the closed form of diarylethene **54** to undergo significant photochemical reaction, which is related to data erasing. For future practical application of this two-photon readout system, the relatively expensive femtosecond Ti:sapphire laser can be replaced with cheaper nanosecond laser diodes ($\lambda = 785$ nm) with comparable output laser intensity [94].

Figure 28e suggests that fluorene **17** indeed underwent 2PA as evidenced by the slope of the plot of fluorescence emission intensity vs. several pump powers. It is particularly noteworthy that with this method of two-photon

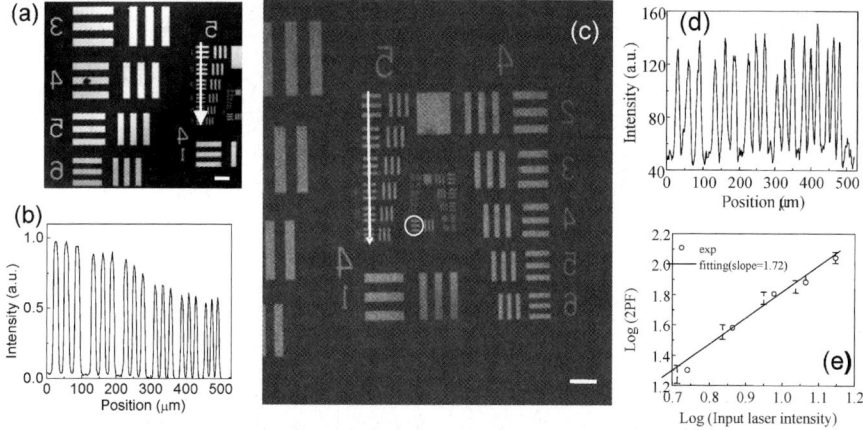

Fig. 28 a Single-photon fluorescence readout of data recorded by single-photon writing (*scale bar*: 100 μm); **b** intensity profile (the direction is shown by the *arrows*) of (**a**); **c** two-photon fluorescence readout of data recorded by single-photon writing (*scale bar*: 100 μm); **d** intensity profile (the direction is shown by the *arrows*) of (**c**); **e** quadratic dependence of up-converted fluorescence of fluorene **17** on the input intensity. The smallest readout pattern achieved in this system was ∼3.5 μm

read-out, data was read out up to 10^4 times with only a negligible decrease in the contrast and fluorescence intensity of the original image. Therefore this method may be considered to be practically non-destructive. One can easily imagine binary data storage based on this system in which a fluorescence intensity threshold is set (e.g., in Fig. 28) with anything below this intensity a "0" and anything above it a "1". Moreover, thirty-eight layers of recorded data was read out without cross talk between adjacent layers, as shown in Fig. 29, demonstrating the potential of this methodology.

Due to the small δ_{2PA} of the closed form of **54**, the two-photon writing was accomplished by using high incident intensity at the focus (90 mW). A rectangular pattern (consisting of 445×345 bits) was recorded in the storage medium, by repeated scanning of the laser beam across the rectangular area for 1.2 s/scan. The total exposure time of each bit was estimated to be about 2 ms. Data was then read by the same two-photon fluorescence microscopy method described above using a low incident intensity of 7 mW (Fig. 30b) and essentially remains unchanged during the readout process. Furthermore, the quadratic relationship between fluorescence emission intensity and incident power provides strong evidence that the data readout in Fig. 30b is, indeed, a result of two-photon induced fluorescence (Fig. 30a).

Another interesting system that was studied for two-photon 3D data storage in our group is based on the modulation of fluorescent properties of 2PA chromophores by protonation [95]. For example, due to differences in basicity (pK_b), fluorene **3** undergoes selective, stepwise protonation, first by proto-

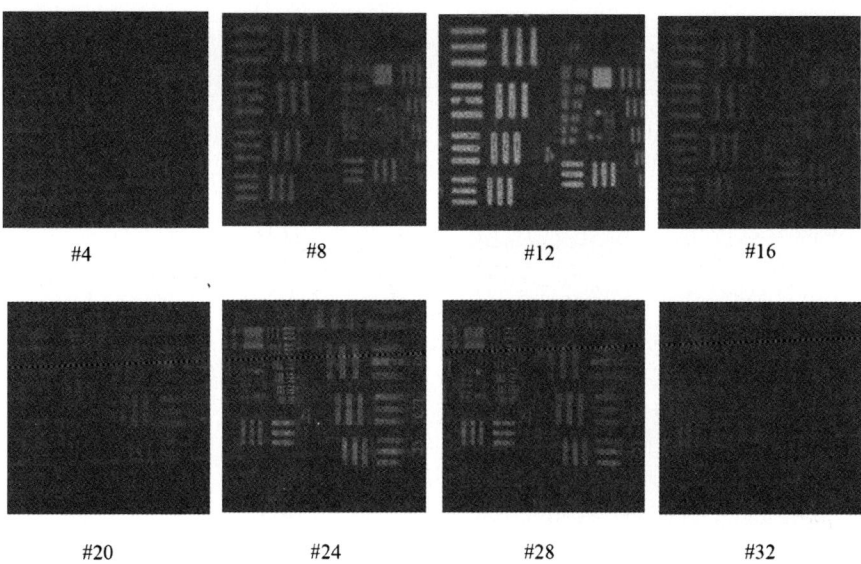

Fig. 29 Two-photon readout of 38 consecutive layers. The layer intervals were 5 μm

nation of the benzothiazolyl nitrogen (pK$_b$ = 13) and then the triarylamino nitrogen (pK$_b$ = 19). This leads to a mixture of three species in Fig. 31 (**3**, **3a**, and **3b**), each with distinct UV-visible absorption and fluorescence emission properties. Time-dependent UV-visible absorption spectra for a solution containing **3** and the photoacid generator (PAG) CD1010 (a triarylsulfonium salt) illustrate this nicely, as shown in Fig. 31.

Upon irradiation with broadband UV light (300–400 nm, 0.57 mW/cm^2), **3** undergoes protonation, resulting in the formation of **3a**, whose absorption spectrum is red-shifted by about 100 nm relative to that of **3**. The conversion of the neutral fluorophore **3** at early irradiation times (10 s) results in decreasing absorbance at its maximum at 390 nm and increasing absorbance at 500 nm upon generation of the protonated form, **3a**. The red shift was expected since fluorene **3** is of an electron donor-π-acceptor construct and protonation of the benzothiazolyl acceptor increases the electron deficiency of this group, affording a greater dipole moment and polarizability. When **3a** undergoes protonation, a new absorption that is blue-shifted relative to both **3** and **3a** was observed due to the fact that the once electron-donating diphenylamino group in **3** and **3a** was converted to an electron-accepting moiety (quaternary ammonium salt) in **3b**. Changes in the fluorescence emission spectra corresponded to the observed changes in the absorption spectra. Protonation of **3** also resulted in a reduction of its fluorescence emission, while emission at longer wavelengths was observed due to excitation of the longer wavelength absorbing monoprotonated **3a**. The fluorescence emission intensity at ~490 nm (390 nm excitation wavelength) decreases with irradi-

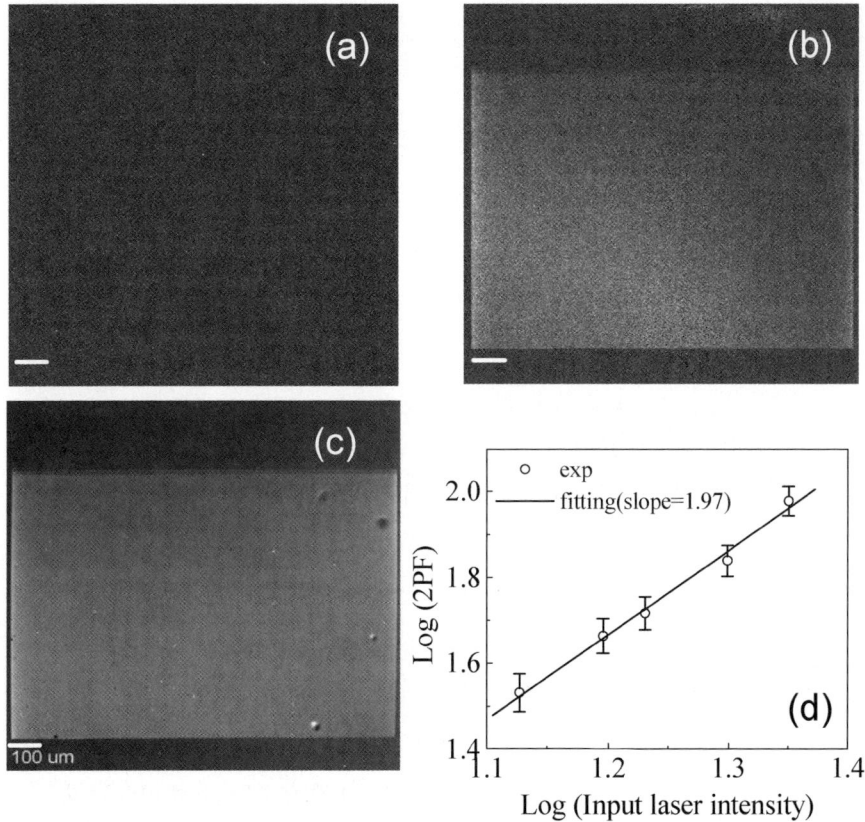

Fig. 30 Two-photon fluorescence image of storage medium before (**a**) and after (**b**) two-photon recording, **c** DIC readout of the storage data after two-photon recording, and **d** input intensity dependent up-converted fluorescence of fluorene **17**

ation while, at early exposure times, emission at ca. 625 nm appears, which then blue shifts upon further protonation to **3b**. The emission at 625 nm is from monoprotonated **3a** upon excitation at 500 nm. Eventually, diprotonation results in a relatively weak, blue-shifted emission at ca. 445 nm (from **3b**). Thus, in addition to observing fluorescence quenching at ca. 490 nm, fluorescence enhancement (creation) at longer wavelengths (ca. 625 nm) is observed upon short irradiation times.

This behavior facilitates two-channel fluorescence imaging, resulting in contrast due to fluorescence quenching at the shorter wavelengths (λ_{em} of **3** from 425 to 620 nm) and fluorescence enhancement at longer wavelengths (λ_{em} of **3a** from 520 to 700 nm) as presented in Fig. 32.

Both writing and recording were accomplished by two-photon excitation of a spin-coated film containing fluorene **3**, the photoacid generator, and polystyrene or, alternatively, in which writing was accomplished by xy scans

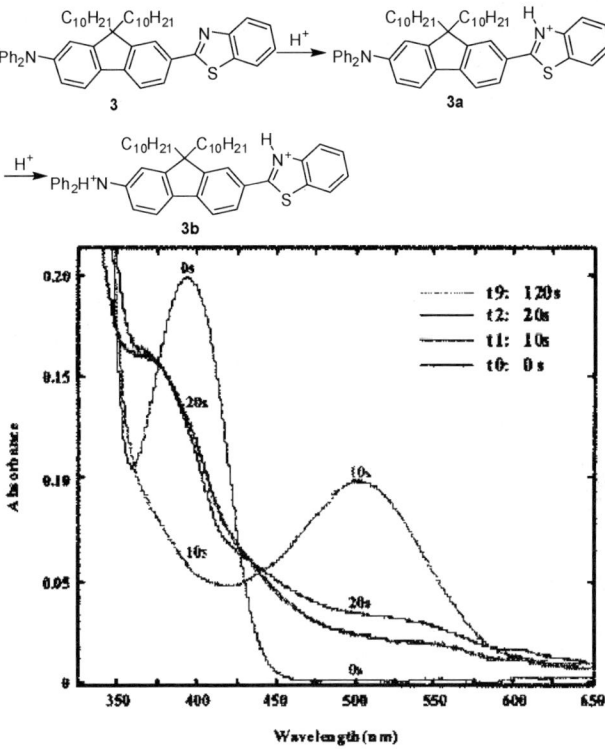

Fig. 31 Time-dependent UV-visible absorption spectra of the photolysis of **3** and photo-acid generator at photolysis times from $t = 0$ to 120 s

at 740 nm (115 fs, 76 MHz). The written image was read by two-photon fluorescence imaging at 800 nm (115 fs, 76 MHz), as shown in Fig. 33. The behavior and relative stability of **3** makes this compound a good candidate for WORM three-dimensional memory systems with writing and reading accomplished via two-photon fluorescence imaging.

4.3
Two-photon Photodynamic Therapy

Photodynamic therapy (PDT) is a treatment that kills cells by irradiating a photosensitizer or photosensitizing agent by a particular type of light. When photosensitizers are exposed to a specific wavelength of light, they produce molecular singlet oxygen, O_2 ($^1\Delta_g$), that destroys nearby cells. The mechanism of production of O_2 ($^1\Delta_g$) is shown in Fig. 34. O_2 ($^1\Delta_g$) can be generated by excitation of a photosensitizer (PS) molecule under linear (one-photon) or nonlinear (two-photon) excitation followed by intermolecular energy transfer to triplet molecular oxygen, O_2 ($^3\Sigma_g^-$) [96]. Relative to sin-

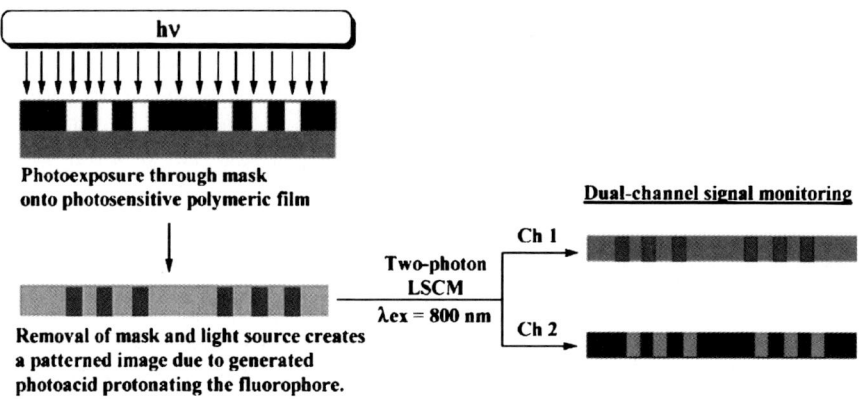

Fig. 32 Diagram of image formation within a photosensitive polymeric film containing PAG, and acid-sensitive fluorophore, allowing two-photon-induced, dual-channel fluorescence imaging

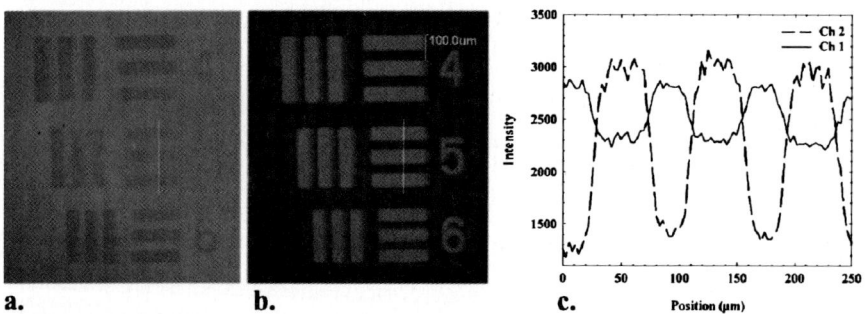

Fig. 33 Two-photon fluorescent images of photosensitive films developed (via 350-nm broadband exposure, 4.4 mW/cm²) using an Air Force resolution target mask. **a** Image recorded by channel 1, **b** image recorded by channel 2, and **c** fluorescence intensity by scanning an *xy* line across one set of three-membered elements (*line* across set 5)

gle photon excitation, two photon excitation again possesses the advantage of higher spatial resolution, and the longer excitation wavelength allowing for deeper penetration into biological media. This is an area that has attracted increased attention. Some recent examples of two-photon photosensitized production of singlet oxygen are difuranonaphthalenes [97], porphyrin [98, 99], phenylene-vinylene-based molecules [100], and fluorene [75] derivatives. However, these PS either possess low efficiency of 2PA or poor compatibility with the biological environment. In general, it is difficult to design a molecule with high singlet oxygen quantum yield generation, a large 2PA cross section, and good biocompatibility. Significant efforts are still needed to attain these goals. In our work, we found that certain fluorene derivatives are promising for this application. The results of singlet oxygen quantum yields of four

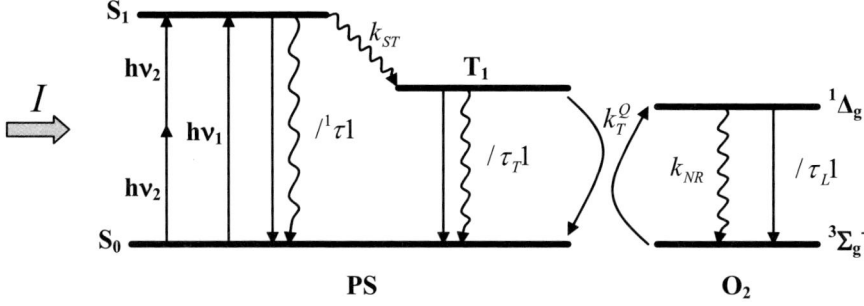

Fig. 34 Photosensitized singlet oxygen production: $1/^1\tau$ is the general (radiative and non-radiative) rate constant of the transition $S_1 \rightarrow S_0$; k_{ST} is the rate constant of singlet-triplet conversion; τ_T is the lifetime of the triplet, T_1, electronic state of PS; k_T^Q is the second-order rate constant of singlet oxygen quenching of the T_1 state of PS; τ_L and k_{NR} are the radiative lifetime and rate constant of all intramolecular nonradiative energy relaxation processes of O_2 ($^1\Delta_g$)

fluorene derivatives are listed in Table 5 as well as those of three standard photosensitizers (Rose Bengal, methylene blue, and hematoporphyrin HP). The structures of the PS are shown in Fig. 35. The results are encouraging, prompting us to further pursue two-photon induced singlet oxygen generation of the most promising candidates. We have developed two methods for singlet oxygen determination under near-IR two-photon excitation, and the efficiency of these compoundsin terms of quantum yield was found to be very high.

For the development of the measurement methodology, it is anticipated that the same mechanism of O_2 ($^1\Delta_g$) generation occurs under both one- and two-photon excitation. This expectation is reasonable due to the very short duration of the femtosecond laser pulse used for two-photon excitation. Thus,

Table 5 Singlet oxygen quantum yields (Φ_Δ) of different PS in ethanol obtained by chemical quenching of DPBF and published values

Compound	λ_{max} (nm)	Φ_Δ measured (DPBF)	Published values of Φ_Δ obtained by		
			DPBF method	SOLM (±0.05)	Other methods
Rose bengal	560	0.80 ± 0.08	$0.79 \pm 0.06/0.86$	0.79 ± 0.05	0.68
Methylene blue	660	0.50 ± 0.05	0.49/0.49	0.42	0.52/0.50
HP	400	0.55 ± 0.05			0.57 ± 0.04
28	344	0.35 ± 0.03			
29	362	0.53 ± 0.05			
55	357	0.53 ± 0.05			
56	314	0.75 ± 0.08			

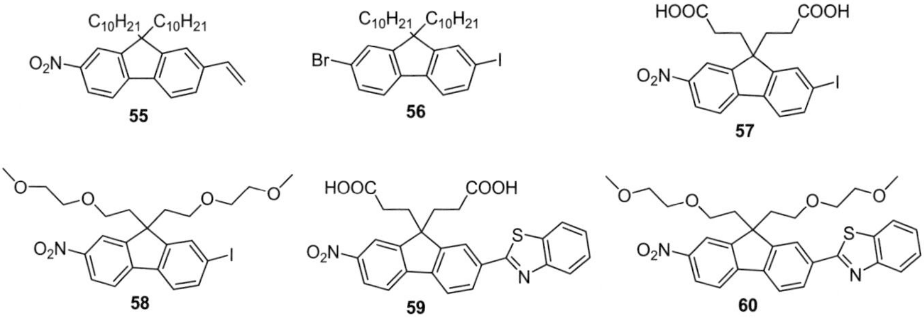

Fig. 35 Molecular structures of singlet oxygen generating fluorene derivatives

it can be expected that all photochemical processes start from the same relaxed first electronic excited state of the PS. Our first method for two-photon singlet oxygen quantum yield measurement was based on an established photochemical method involving the reaction of O_2 ($^1\Delta_g$) with a singlet oxygen-sensitive compound (quencher) [101] in combination with accurate, time-dependent spectrofluorimetric determination of quencher concentration. 1,3-Diphenylisobenzofuran (DPBF) was used as an efficient chemical quencher.

The DPBF-photooxidative product quantum yield Φ_{DPBFO_2}, and singlet oxygen quantum yield of the PS, Φ_Δ^{2PA}, are related according to the literature as [102]:

$$\frac{1}{\Phi_{DPBFO_2}} = \frac{1}{\Phi_\Delta^{2PA}} + \frac{\beta}{\Phi_\Delta^{2PA} \cdot C_{DPBF}}, \qquad (16)$$

where $\beta = k_d/k_a$ is constant for the DPBF decomposition in certain solvents [102]. Equation 16 was used to obtain the value of Φ_Δ and β from the Stern–Volmer plot, $(1/\Phi_{DPBFO_2})$ versus $(1/C_{DPBF})$. With this method, a two-photon generated singlet oxygen quantum yield of 0.4 ± 0.1 was determined for compound **29** in ethanol at excitation wavelength 720 nm, which coincides with the corresponding singlet oxygen quantum yield determined under one-photon excitation of $\Phi_\Delta \approx 0.53 \pm 0.05$, as listed in Table 5.

Although the above indirect method provided relatively accurate results, a new method of direct measurement of the IR phosphorescence of O_2 ($^1\Delta_g$) at 1270 nm was expected to improve the measurement [103]. The quantum yield of two-photon singlet oxygen generation, $^{2PA}\Phi_\Delta$, was calculated from Eq. 17:

$$\Phi_\Delta^{2PA} = \Phi_{\Delta R}^{2PA} \frac{I_\Delta \delta_R^{2PA}}{I_{\Delta R} \delta^{2PA}}, \qquad (17)$$

where I_Δ, $I_{\Delta R}$, δ^{2PA} and δ_R^{2PA} are the average steady-state phosphorescence emissions of singlet oxygen from the PS solutions and 2PA cross sections of the sample and reference PS at the excitation wavelength ($\lambda_{exc} = 775$ nm), respectively, and $\Phi_{\Delta R}^{2PA}$ is the quantum yield of O_2 ($^1\Delta_g$) sensitization under two-photon excitation of the reference PS.

The typical phosphorescence signal of O_2 ($^1\Delta_g$) produced under single-photon excitation of PS **60** in ACN is shown in Fig. 36. These signals are proportional to I_Δ and were used for the determination of Φ_Δ^{2PA}. The value of I_Δ exhibited a linear dependence on the square of excitation power for all the investigated PS, indicating pure 2PA processes and low quenching efficiency with rate constants k^Q and k_Δ^Q. 2PA cross sections, δ^{2PA}, of PS **57–60** in ACN were measured at 775 nm by the open aperture Z-scan method and are presented in Table 6. Relatively low values of δ^{2PA} were obtained for PS **57** and **58** at 775 nm. This wavelength corresponds to the long wavelength edge of their main absorption band. Although 775 nm is not optimal for two-photon photosensitization, this wavelength was used since it is the fundamental output of the Clark-MXR Ti:sapphire laser and possessed sufficient intensity for 2PA-induced singlet oxygen quantum yield determination. Under excitation in two-photon allowed electronic states, fluorene derivatives exhibit $\delta^{2PA} \geq 200$–300 GM and can be successfully utilized. The quantum yields, Φ_Δ^{2PA}, were determined from Eq. 14 for PS **1–4** in ACN and are presented in Table 6. As can be seen from these data, PS **57–58** all possessed relatively high two-photon quantum yields, $\Phi_\Delta^{2PA} \approx 0.3$–$0.45$, which are nearly equal to half of the corresponding Φ_Δ values obtained under linear, one-photon, excitation. This indicates that the processes of singlet oxygen production for all the

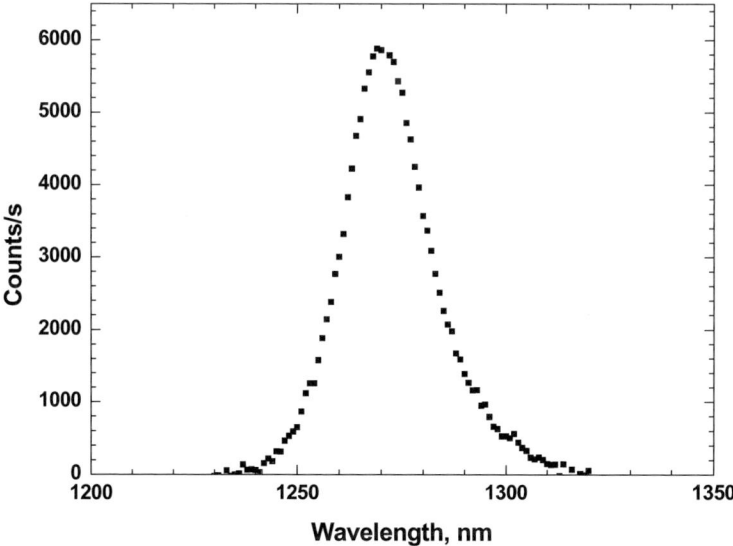

Fig. 36 Phosphorescence signal of singlet oxygen produced by PS **60** in ACN under steady-state one-photon excitation at $^{abs}\lambda_{max}$

Table 6 Two-photon absorption cross sections, δ_{2PA}, and quantum yields of singlet oxygen generation, Φ_Δ^{2PA}, under single and two-photon excitation of **57–60** at 775 nm in ACN

Compound	λ_{max}	Φ_Δ	δ^{2PA}, GM	Φ_Δ^{2PA}
57	340 ± 1	0.65 ± 0.07	9 ± 4	0.4 ± 0.2
58	342 ± 1	0.74 ± 0.08	10 ± 4	0.3 ± 0.15
59	362 ± 1	0.93 ± 0.1	50 ± 15	0.35 ± 0.1
60	364 ± 1	0.92 ± 0.1	60 ± 20	0.45 ± 0.15

fluorenyl PS start from the same vibronic levels of S_1 and are nearly independent of the type of excitation.

4.4
Two-photon 3D Microfabrication

Emerging device technologies such as microelectromechanical systems and integrated sensors are placing increased demands on the development of 3D materials processing and fabrication techniques. Traditional 3D microfabrication techniques for microscale free-form fabrication include microstereolithography, electrochemical fabrication (EFAB), microphotoforming, spatial forming, microtransfer molding, localized electrochemical deposition, etc [104]. With these techniques, 3D objects were successfully fabricated from the various materials, including polymer, ceramic, metal, etc. Most of these techniques build 3D structures using layer-by-layer photopolymerization. For example, in a typical microstereolithographic process a CAD model of the desired object is first generated, and then the 3D model is sliced into a series of closely spaced horizontal planes and converted into computer-executable codes. Controlled by these codes, the desired polymer object is built by computer-directed laser scanning on the surface of a photocurable liquid resin in a layer-by-layer additive fashion. In contrast, due to the characteristic 3D spatial resolution of the simultaneous 2PA process, polymerization induced by 2PA can be harnessed to directly write 3D objects in photocurable materials. This is facilitated by the unique properties associated with simultaneous absorption of two-photon relative to single-photon mediated processes.

While a more detailed review of two-photon 3D microfabrication can be found in another chapter of this issue, this chapter will concentrate on our work on characterization and application of commercial photoinitiators. The development of better 2PA photoinitiators would be expected to facilitate 3D photopolymerization technologies. Efficient 2PA compounds based on phenylethenyl constructs bearing electron-donating and/or electron-withdrawing moieties have been reported [105–107]. Among these are electron-rich derivatives that have been found to undergo a presumed two-photon induced electron transfer to acrylate monomers or a proposed fluorescence energy transfer to a photoinitiator of polymerization. The reportedly efficient two-photon photoinitiators, although more photosensitive than previously studied UV photoinitiators, are not commercially available and require rather involved syntheses. As such, many 2PA-induced polymerizations directed towards micro- and optical-component fabrication continue to employ conventional, commercially available UV-visible photoinitiators. There have been a limited number of reports of two-photon photopolymerization of commercial acrylate monomer systems, pre-formulated with UV photoinitiators [85, 108]. Additionally, previous 2PA microfabrication efforts using commercial UV resins and photopolymerizable systems reveal little information pertaining to the photophysical properties of the photoinitiators used. As these initiators are in current and continual use, information on their 2PA properties may facilitate optimization of 2PA-induced polymerization conditions. Given this, the 2PA photophysical properties of several common photoinitiators were measured [58]. Their structures are shown in Fig. 37.

In the linear absorption measurements of these photolabile compounds, there is a concern that during the measurement of the 2PA cross sections the photoinitiators may photodecompose, which would in turn affect the observed cross section, especially when the Z-scan technique is used. The Z-

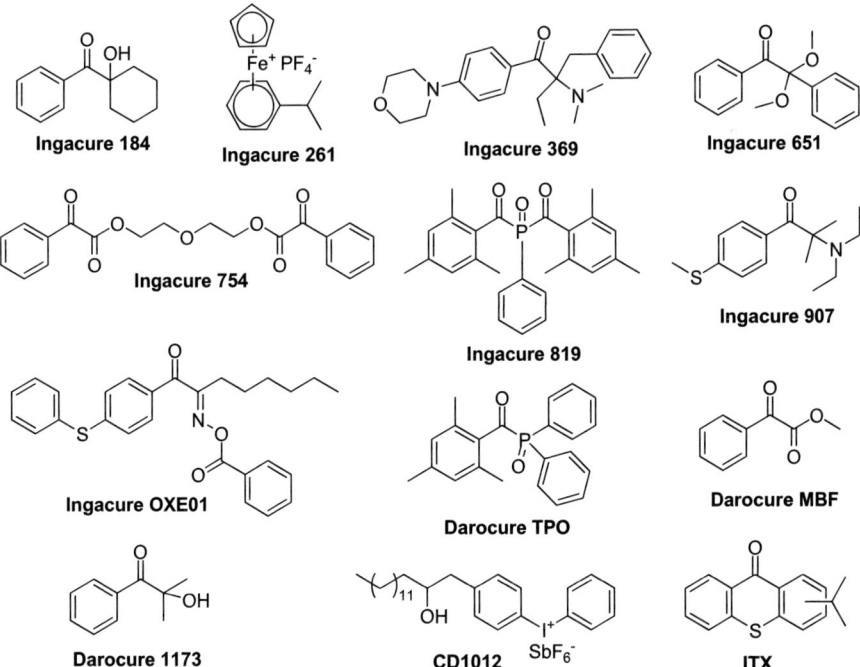

Fig. 37 Structures of commercial photoinitiators

scan technique requires higher irradiances (50–$400\,\mathrm{GW/cm^2}$), consequently resulting in higher probabilities of photodegradation compared to the use of the white-light continuum (WLC) method ($<50\,\mathrm{GW/cm^2}$). Furthermore, in the WLC method, there is no degenerate 2PA of the strong pump beam, only simultaneous non-degenerate 2PA of the pump and probe beams. Since the monitored change in transmittance is of the *weak* probe beam, this guarantees that the population promoted into the excited state is small. Typical 2PA spectra of three photoinitiators ITX, Irgacure 369 and Irgacure OXE01 using the WLC method are presented in Fig. 38.

For the results from the Z-scan method, the effect of the potential photodegradation was carefully evaluated by performing the measurement on solutions in an enclosed cuvette as well as in a flow cell (both with 1 mm path length). In the flow cell geometry, the flow rate was set to assure that a fresh volume of solution was measured with nearly every laser pulse. The nonlinear signal from the flow cell geometry for Irgacure OXE01 was 3.5 times larger than that of the stationary solution, indicating that photodegradation can interfere with the measurement of the 2PA cross section if precautions are not taken to avoid this. This also suggests that Irgacure OXE01 should be an effective 2PA photoinitiator since it readily undergoes photolysis upon near-IR two-photon excitation. Listed in Table 7 are two-photon absorption cross sections of a number of commercial photoinitiators by two methods, i.e. by the Z-scan technique using the flow cell and by the femtosecond WLC pump–probe method.

In addition to these examples, an electron-transfer free radical photoinitiator H-Nu 470 (5,7-diiodo-3-butoxy-6-fluorone) has been also successfully used for 3D microfabrication by near-IR two-photon induced polymerization

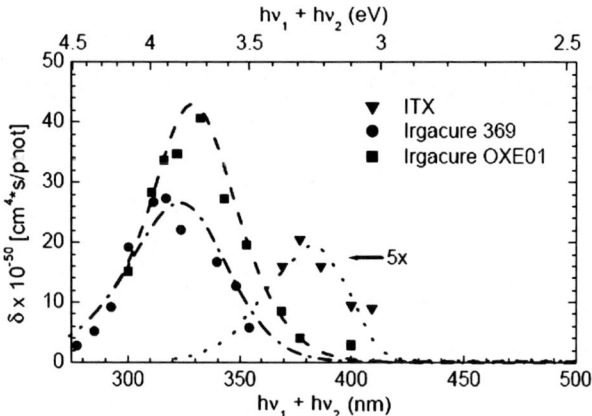

Fig. 38 2PA spectra via the WLC method. ITX spectrum has been enlarged ($5\times$) for ease of viewing. *Lines* indicate UV-visible linear absorption spectra: ITX (*dot*), Irgacure 369 (*dash-dot*), and Irgacure OXE01 (*dash*)

Table 7 Peak 2PA cross sections for photoinitiators using Z-scan and WLC methods

Compound	λ_{max}	Z-scan		WLC-2PA	
		$\lambda_{max}^{(2)}$	δ_{meas}	$\lambda_{max}^{(2)}$	δ_{meas}
Irgacure 184	246	265	23	250	< 20
Irgacure 261	242	265	< 20	250	< 20
Irgacure 369	324	335	7	318	27
Irgacure 651	254	265	28	250	< 20
Irgacure 754	253	265	21	250	10
Irgacure 819	295	300	< 4	300	< 5
Irgacure 907	306	300	4	300	< 5
Irgacure OXE01	328	330	31	330	38
Darocure TPO	299	300	< 4	300	< 5
Darocure MBF	255	265	27	250	< 20
Darocure 1173	244	265	< 20	250	< 20
CD 1012	247	265	16	273	14
ITX	382	38	5	377	4

of (meth)acrylate monomers. The polymerization was initiated at 775 nm via direct excitation of H-Nu 470 in the presence of an arylamine, and (meth)acrylate monomer. The formation of polymeric microstructures with a variety of dimensions was accomplished. In simple line scans, line widths were reproducibly produced with uniform line widths ranging from 7 to 15 μm, spaced 20–50 μm apart. Microstructures were readily examined by optical reflection microscopy. Figure 2 depicts the microstructure formed in which the line width is 9 μm with relatively uniform 50 μm line spacing. Elec-

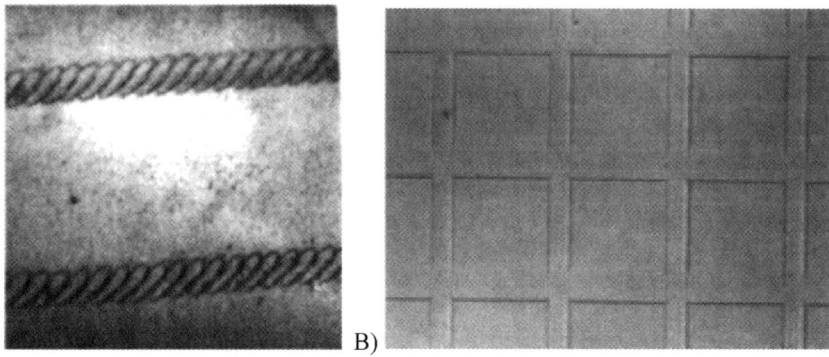

Fig. 39 Micrograph of polymerized uniform submicrostructure with 9 μm line width and 50 μm line spacing. The structure was written by two-photon initiated electron-transfer free radical polymerization of diacrylate monomer Sartomer SR 349 at 775 nm via direct excitation of dye 5,7-diiodo-3-butoxy-6-fluorone(H-Nu 470) for (**A**) and dye 3 for (**B**) in the presence of N,N-dimethyl-2,6-diisopropylaniline

tron transfer from the aromatic amine (*N*,*N*-dimethyl-2,6-diisopropylaniline) to the fluorone derivative, followed by proton transfer from the amine to the fluorone, resulted in formation of an arylamine bearing a free radical localized on the α-methylene carbon. This free radical species then initiated polymerization of (meth)acrylate derivatives. A diaryliodonium salt can be added to accelerate the rate of polymerization [108].

As another example, fluorene 3 was found to be an effective initiator for an acrylate (SR349) polymerization via 2PE at 775 nm, presumably by means of an electron transfer process. The resulting microstructure had 18 μm line widths of uniform spacing as shown in Fig. 39B [85].

5
Conclusion

The last decade of the 20th century and the first decade of the 21st century has witnessed growing interest in two-photon, nonlinear optical materials and applications. The highly advantageous properties associated with long wavelength, two-photon absorption will continue to inspire this activity. Applications that employ two-photon absorption are making their way from the laboratory to the market and will continue to do so. As part of our efforts in this field, appropriately functionalized fluorene derivates have been prepared that possess high two-photon absorptivity and high photostability under both one-photon and two-photon excitation. Furthermore, derivatives can be molecularly engineered to exhibit high fluorescence quantum yield or high rates of intersystem crossing to the triplet state and subsequent singlet oxygen generation (singlet oxygen photosensitizers). Biocompatibility can be readily imparted to the fluorine derivatives by introduction of hydrophilic substituents at the 9-position, providing outstanding two-photon fluorescent probes for organelle-specific bioimaging and photodynamic therapy. The inherent three-dimensional spatial resolution inherent in 2PE, along with the use of near-IR light, affords unprecedented opportunities in the development of photonic materials and devices for a number of technologically important applications such as high density 3D data storage and 3D lithography and microfabrication.

As more sensitive materials with specifically tailored properties and sound theoretical method development progress, scientists and engineers should have a predictive capability to rationally design tailor-made multiphoton absorbing materials for particular applications. This should help propel the numerous applications of two-photon and higher order multiphoton absorbing materials described above, along with emerging areas such as optical signal processing and quantum computing. The continued dependence on the interdisciplinary expertise of materials chemists, spectroscopists, theorists, and engineers makes the field of multiphoton based research and device development an exciting and virile field for the next generation of researchers.

References

1. Goppert-Mayer M (1931) Annal Physik 9:273
2. Parthenopoulos DA, Rentzepis PM (1989) Science 245:843
3. Denk W, Strickler JH, Webb WW (1990) Science 248:73
4. He GS, Xu GC, Prasad PN, Reinhardt BA, Bhatt JC, Dillard AG (1995) Opt Lett 20:435
5. Narang U, Zhao CF, Bhawalkar JD, Bright FV, Prasad PN (1996) J Phys Chem 100:4521
6. Reinhardt BA, Brott LL, Clarson SJ, Dillard AG, Bhatt JC, Kannan R, Yuan L, He GS Prasad PN (1998) Chem Mater 10:1863
7. Albota M, Beljonne D, Bredas J-L, Ehrlich JE, Fu J-Y, Heikal AA, Hess SE, Kogej T, Levin MD, Marder SR, McCord-Maughon D, Perry JW, Rockel H, Rumi M, Subramaniam G, Webb WW, Wu X-L, Xu C (1998) Science 281:1653
8. Kogej T, Beljonne D, Meyers F, Perry JW, Marder SR, Bredas JL (1998) Chem Phys Lett 298:1
9. Perry JW, Barlow S, Ehrlich JE, Heikal AA, Hu ZY, Lee IY, Mansour K, Marder SR, Rockel H, Rumi M, Thayumanavan S, Wu XL (1999) MCLC S&T, Sect B: Nonlinear Opt 21:225
10. Kannan R, He GS, Yuan L, Xu F, Prasad PN, Dombroskie AG, Reinhardt BA, Baur JW, Vaia RA, Tan L-S (2001) Chem Mater 13:1896
11. He GS, Swiatkiewicz J, Jiang Y, Prasad PN, Reinhardt BA, Tan L-S, Kannan R (2000) J Phys Chem A 104:4805
12. Belfield KD, Schafer KJ, Mourad W, Reinhardt BA (2000) J Org Chem 65:4475
13. Lakowicz JR (1999) Principles of Fluorescence Spectroscopy. Kluwer Academic/Plenum Publisher, New York
14. Baur JW, Alexander MD, Banach M, Denny LR, Reinhardt BA, Vaia RA, Fleitz PA, Kirkpatrick SM (1999) Chem Mater 11:2899
15. Belfield KD, Bondar MV, Kachkovsky OD, Przhonska OV, Yao S (2007) J Luminesc 126:14
16. Parker CA (1968) Photoluminescence of Solutions: with applications to photochemistry and analytical chemistry. Elsevier, New York
17. Lahmani F, Breheret E, Benoist d'Azy O, Zehnacker-Rentien A, Delouis JF (1995) J Photochem Photobiol A: Chem 89:191
18. Barzoukas M, Runser C, Fort A, Blanchard-Desce M (1996) Chem Phys Lett 257:531
19. Bosch LI, Mahon MF, James TD (2004) Tetrahedron Lett 45:2859
20. Belfield KD, Bondar MV, Przhonska OV, Schafer KJ (2002) J Fluoresc 12:449
21. Belfield KD, Bondar MV, Przhonska OV, Schafer KJ, Mourad W (2002) J Luminesc 97:141
22. Belfield KD, Bondar MV, Cohanoschi I, Hernandez FE, Kachkovsky OD, Przhonska OV, Yao S (2005) Appl Opt 44:7232
23. Fu J, Przhonska OV, Padilha LA, Hagan DJ, Van Stryland EW, Belfield KD, Bondar MV, Slominsky YL, Kachkovsky AD (2006) Chem Phys 321:257
24. Strickler SJ, Berg RA (1962) J Chem Phys 37:814
25. Belfield KD, Bondar MV, Hernandez FE, Morales AR, Przhonska OV, Schafer KJ (2004) Appl Opt 43:6339

26. Sheik-Bahae M, Said AA, Wei TH, Hagan DJ, Van Stryland EW (1990) Quan Electron, IEEE J 26:760
27. Ippen EP, Shank CV (1977) In: Shapiro SL (ed) Topics in Applied Physics. Springer-Verlag, London, p 83
28. Przhonska OV, Bondar MV, Slominsky YL (2001) Sci Appl Photo 43:71
29. Reindl S, Penzkofer A (1998) Chem Phys 230:83
30. Cronstrand P, Luo Y, Agren H (2002) J Chem Phys 117:11102
31. Cronstrand P, Luo Y, Agren H (2002) Chem Phys Lett 352:262
32. Lessing HE, Von Jena A (1978) Chem Phys Lett 59:249
33. Penzkofer A, Wiedmann J (1980) Opt Comm 35:81
34. Myslinski P, Koningstein JA, Shen Y (1992) J Chem Phys 96:8691
35. He GS, Lin T-C, Prasad PN, Kannan R, Vaia RA, Tan L-S (2002) J Phys Chem B 106:11081
36. Kohler RH, Cao J (1997) Science 276:2039
37. Bestvater F, Spiess E, Stobrawa G, Hacker M, Feurer T, Porwol T, Berchner-Pfannschmidt U, Wotzlaw C, Acker H (2002) J Microsc 208:108
38. Miller MJ, Wei SH, Parker I, Cahalan MD (2002) Science 296:1869
39. Ehrlich JE, Wu XL, Lee IYS, Hu ZY, Rockel H, Marder SR, Perry JW (1997) Opt Lett 22:1843
40. Parthenopoulos DA, Rentzepis PM (1990) J Appl Phys 68:5814
41. Strickler JH, Webb WW (1991) Opt Lett 16:1780
42. Kawata S, Sun H-B, Tanaka T, Takada K (2001) Nature 412:697
43. Bhawalkar JD, He GS, Prasad PN (1996) Rep Prog Phys 59:1041
44. Dick B, Hochstrasser RM, Trommsdorff HP (1987) In: Chemla DS, Zyss J (eds) Nonlinear Optical Properties of Organic Molecules and Crystals. Academic, New York, p 159, p 503
45. Loudon R (1983) The Quantum Theory of Light. Oxford University Press, Oxford
46. Zojer E, Beljonne D, Kogej T, Vogel H, Marder SR, Perry JW, Bredas JL (2002) J Chem Phys 116:3646
47. Beljonne D, Wenseleers W, Zojer E, Shuai Z, Vogel H, Pond SJK, Perry JW, Marder SR, Bredas J-L (2002) Adv Funct Mater 12:631
48. Meyers F, Marder SR, Perry JW (1998) In: Interrante LV, Hampden-Smith MJ (eds) Chemistry of Advanced Materials: An Overview. Wiley-Interscience, New York, p 207
49. Karna SP, Yeates AT (1996) In: Karna SP, Yeates AT (eds) Nonlinear Optical Materials: Theory and Modeling. American Chemical Society Publication, Washington DC, p 1
50. Bredas JL (1993) In: Zerbi G (ed) Organic Materials for Photonics: Science and Technology. North-Holland, Amsterdam, p 127
51. Orr BJ, Ward JF (1971) Mol Phys 20:513
52. Boyd RW (1992) Nonlinear Optics. Academic Press, San Diego
53. Yao S, Belfield KD (2005) J Org Chem 70:5126
54. Beljonne D, Bredas JL, Cha M, Torruellas WE, Stegeman GI, Hofstraat JW, Horsthuis WHG, Mohlmann GR (1995) J Chem Phys 103:7834
55. Rumi M, Ehrlich JE, Heikal AA, Perry JW, Barlow S, Hu Z, McCord-Maughon D, Parker TC, Roeckel H, Thayumanavan S, Marder SR, Beljonne D, Bredas J-L (2000) J Am Chem Soc 122:9500
56. Hales JM, Hagan DJ, Van Stryland EW, Schafer KJ, Morales AR, Belfield KD, Pacher P, Kwon O, Zojer E, Bredas JL (2004) J Chem Phys 121:3152

57. Hales JM (2004) Chemical Structure-Nonlinear Optical Property Relationships for A Series of Two-Photon Absorbing Fluorene Molecules. CREOL and School of Optics, University of Central Florida, Orlando, PhD dissertation
58. Schafer KJ, Hales JM, Balu M, Belfield KD, Van Stryland EW, Hagan DJ (2004) J Photochem Photobiol A: Chem 162:497
59. Belfield KD, Morales AR, Hales JM, Hagan DJ, Van Stryland EW, Chapela VM, Percino J (2004) Chem Mater 16:2267
60. Xu C, Webb WW (1996) J Opt Soc Am B: Opt Phys 13:481
61. Negres RA, Hales JM, Kobyakov A, Hagan DJ, Van Stryland EW (2002) Quan Electron IEEE J 38:1205
62. Peticolas WL (1967) Ann Rev Phys Chem 18:233
63. Belfield KD, Yao S, Hales JM, Bondar MV, Hagan DJ, Van Stryland EW (2004) PMSE Prepr 91:340
64. Kuzyk MG (2003) J Chem Phys 119:8327
65. Volkmer A, Subramaniam V, Birch DJS, Jovin TM (2000) Biophys J 78:1589
66. Strehmel B, Sarker AM, Detert H (2003) ChemPhysChem 4:249
67. Zojer E, Wenseleers W, Pacher P, Barlow S, Halik M, Grasso C, Perry JW, Marder SR, Bredas J-L (2004) J Phys Chem B 108:8641
68. Wood PD, Johnston LJ (1998) J Phys Chem A 102:5585
69. Turro NJ (1965) Molecular Photochemistry. WA Benjamin, New York
70. Belfield KD, Bondar MV, Przhonska OV, Schafer KJ (2004) J Photochem Photobiol A: Chem 162:569
71. Belfield KD, Bondar MV, Przhonska OV, Schafer KJ (2004) J Photochem Photobiol A: Chem 162:489
72. Belfield KD, Bondar MV, Liu Y, Przhonska OV (2003) J Phys Org Chem 16:69
73. Corredor CC, Belfield KD, Bondar MV, Przhonska OV, Yao S (2006) J Photochem Photobiol A: Chem 184:105
74. Belfield KD, Bondar MV, Przhonska OV, Schafer KJ (2004) Photochem Photobiol Sci 3:138
75. Belfield KD, Bondar MV, Przhonska OV (2006) J Fluoresc 16:111
76. Xu C, Williams RM, Zipfel W, Webb WW (1996) Bioimaging 4:198
77. Chung S-J, Zheng S, Odani T, Beverina L, Fu J, Padilha LA, Biesso A, Hales JM, Zhan X, Schmidt K, Ye A, Zojer E, Barlow S, Hagan DJ, Van Stryland EW, Yi Y, Shuai Z, Pagani GA, Bredas J-L, Perry JW, Marder SR (2006) J Am Chem Soc 128:14444
78. Schafer-Hales KJ, Belfield KD, Yao S, Frederiksen PK, Hales JM, Kolattukudy PE (2005) J Biomed Opt 10:051402
79. Cheng PC, Pan SJ, Shih A, Kim KS, Liou WS, Park MS (1998) J Microsc 189:199
80. Belfield KD, Morales AR, Schafer-Hales KJ, Yao S (2006) Polym Prepr 47:1006
81. Meltola NJ, Soini AE, Hänninen PE (2004) J Fluoresc 14:129
82. Meltola NJ Wahlroos R Soini AE (2004) J Fluoresc 14:635
83. Esener SC, Walker EP, Zhang Y, Dvornikov AS, Rentzepis PM (2003) Proc SPIE-Intern Soc Opt Eng 4988:93
84. Belfield KD, Liu Y, Negres RA, Fan M, Pan G, Hagan DJ, Hernandez FE (2002) Chem Mater 14:3663
85. Belfield KD, Schafer KJ, Liu Y, Liu J, Ren X, Van Stryland EW (2000) J Phys Org Chem 13:837
86. Dürr H, Bouas-Laurent H (1990) Photochromism: Molecules and Systems, Chaps 8–10. Elsevier, New York

87. Yokoyama Y, Kose M (1995) In: Horspool WM, Song P-S (eds) CRC Handbook of Organic Photochemistry and Photobiology. CRC Press, Boca Raton Florida, p 86
88. Corredor CC, Huang Z-L, Belfield KD (2006) Adv Mater 18:2910
89. Irie M (2000) Chem Rev 100:1685
90. Tian H, Yang S (2004) Chem Soc Rev 33:85
91. Kozlov DV, Castellano FN (2004) J Phys Chem A 108:10619
92. Giordano L, Jovin TM, Irie M, Jares-Erijman EA (2002) J Am Chem Soc 124:7481
93. Belfield KD, Wang J (1995) J Polym Sci Part A: Polym Chem 33:1235
94. Shiono T, Itoh T, Nishino S (2005) Jap J Appl Phys Part 1 44:3559
95. Belfield KD, Schafer KJ (2002) Chem Mater 14:3656
96. Schweitzer C, Schmidt R (2003) Chem Rev 103:1685
97. Poulsen TD, Frederiksen PK, Jorgensen M, Mikkelsen KV, Ogilby PR (2001) J Phys Chem A 105:11488
98. Karotki A, Drobizhev M, Kruk M, Spangler C, Nickel E, Mamardashvili N, Rebane A (2003) J Opt Soc Am B: Opt Phys 20:321
99. Frederiksen PK, McIlroy SP, Nielsen CB, Nikolajsen L, Skovsen E, Jorgensen M, Mikkelsen KV, Ogilby PR (2005) J Am Chem Soc 127:255
100. Nielsen CB, Johnsen M, Arnbjerg J, Pittelkow M, McIlroy SP, Ogilby PR, Jorgensen M (2005) J Org Chem 70:7065
101. Spiller W, Kliesch H, Wöhrle D, Hackbarth S, Röder B, Schnurpfeil G (1998) J Porphyrins Phthalocyanines 2:145
102. Foote CS (1979) In: Wasserman HH, Murray RW (eds) Singlet Oxygen. Academic Press, NewYork San Francisco London, p 139
103. Andrasik SJ, Belfield KD, Bondar MV, Hernandez FE, Morales AR, Przhonska OV, Yao S (2007) ChemPhysChem 8:399
104. Varadan V, Jiang X, Varadan VV (2001) Microstereolithography and other Fabrication Techniques for 3D MEMS. John Wiley and Sons, Chichester
105. Cumpston BH, Ananthavel SP, Barlow S, Dyer DL, Ehrlich JE, Erskine LL, Heikal AA, Kuebler SM, Lee IYS, McCord-Maughon D, Qin J, Rockel H, Rumi M, Wu X-L, Marder SR, Perry JW (1999) Nature 398:51
106. Kuebler SM, Braun KL, Zhou W, Cammack JK, Yu T, Ober CK, Marder SR, Perry JW (2003) J Photochem Photobiol A: Chem 158:163
107. Zhou W, Kuebler SM, Braun KL, Yu T, Cammack JK, Ober CK, Perry JW, Marder SR (2002) Science 296:1106
108. Belfield KD, Ren X, Van Stryland EW, Hagan DJ, Dubikovsky V, Miesak EJ (2000) J Am Chem Soc 122:1217

Adv Polym Sci (2008) 213: 157–206
DOI 10.1007/12_2007_122
© Springer-Verlag Berlin Heidelberg
Published online: 26 October 2007

Three-Dimensional Structuring of Resists and Resins by Direct Laser Writing and Holographic Recording

Saulius Juodkazis (✉) · Vygantas Mizeikis (✉) · Hiroaki Misawa (✉)

CREST-JST & Research Institute for Electronic Science,
Hokkaido University, N21W10 CRIS Building, 001-0021 Sapporo, Japan
Saulius@es.hokudai.ac.jp, VM@es.hokudai.ac.jp, Misawa@es.hokudai.ac.jp

Abstract Optical techniques for three-dimensional micro- and nanostructuring of transparent and photo-sensitive materials are reviewed with emphasis on methods of manipulation of the optical field, such as beam focusing, the use of ultrashort pulses, and plasmonic and near-field effects. The linear and nonlinear optical response of materials to classical optical fields as well as exploitation of the advantages of quantum lithography are discussed.

Keywords Laser lithography · Optical nonlinearities · Optical structuring ·
Photopolymerization · Quantum lithography

Abbreviations

1D	One-dimensional
2D	Two-dimensional
3D	Three-dimensional
A_e	Entanglement area
c	Speed of light
CCD	Charge coupled device
cw	Continuous-wave (refers to lasers)
DOE	Diffractive optical element
DLW	Direct laser writing
EBL	Electron-beam lithography
FDTD	Finite-domain time difference calculations
$f_\#$	f-Number
FROG	Frequency-resolved optical gating
FWHM	Full-width at half maximum
GRENOUILLE	GRating-Eliminated No-nonsense Observation of Ultrafast Incident Laser Light E-fields
GM	Göppert-Mayer GM units of TPA cross-section (1 GM = 10^{-50} cm^4 s)
GVD	Group velocity dispersion
E	Young modulus
G	Shear modulus
IR	Infra red
k	Spring constant (SI: N/m)
MBAPB	Hydrophobic dye $C_{40}H_{54}N_2O_2$
MPA	Multiphoton absorption
MEMS	Microelectromechanical systems
NEMS	Nanoelectromechanical systems
NA	Numerical aperture
PhC	Photonic crystal
PBG	Photonic band gap
PSB	Photonic stop gap
PSF	Point spread function

PR	Phase retarder
PZT	Piezo-ceramic transducer
R	Absorption rate
S	Irradiation area
SGH	Second harmonic generation
SEM	Scanning electron microscopy
SPDC	Spontaneous parametric down conversion
T_e	Entanglement time
TFSF	Total-field scattered-field
TPA	Two-photon absorption
SU-8	Brand of commercial photoresist from MicroChem, Newton, MA, USA
UV	Ultra-violet
ϕ	Photon flux density
n_2	Nonlinear index of refraction
λ	Wavelength
μ-TAS	Micro-total analysis systems
ν	Poisson's ratio
σ_{ab}	Cross-section of linear absorption
$\sigma_{2\gamma}$	Cross-section of two-photon absorption
τ_p	Pulse duration at FWHM level (for Gaussian pulses)

1
Introduction

Three-dimensional (3D) structuring of materials allows miniaturization of photonic devices, micro-(nano-)electromechanical systems (MEMS and NEMS), micro-total analysis systems (μ-TAS), and other systems functioning on the micro- and nanoscale. Miniature photonic structures enable practical implementation of near-field manipulation, plasmonics, and photonic band-gap (PBG) materials, also known as photonic crystals (PhC) [1, 2]. In micromechanics, fast response times are possible due to the small dimensions of moving parts. Femtoliter-level sensitivity of μ-TAS devices has been achieved due to minute volumes and cross-sections of channels and reaction chambers, in combination with high resolution and sensitivity of optical confocal microscopy. Progress in all these areas relies on the 3D structuring of bulk and thin-film dielectrics, metals, and organic photosensitive materials.

Currently, a range of techniques are used to obtain 3D micro- and nanostructures. They include self-organization, biomimetics, electron-beam lithography (EBL), ion and particle beam-milling, and optical fabrication. Arguably, the best-known used technique of optical fabrication is optical lithography. This term refers to mask projection lithography and implies replication of a two-dimensional (2D) image by drawing on *lithos* (Greek: stone). Intense development of this technique has occurred mainly due to the demands of microelectronics science and industry. Lithographic fabrication of high resolution masks, combined with thin-film deposition of various materials, can produce multilayered 3D microstructures composed of semi-conductors, metals, and organics as well as non-organic dielectrics, which are typically built in a layer-by layer manner from finely structured 2D layers. Despite the great successes and achievements of this approach in the field of microelectronics, the novel 3D microstructures mentioned above require fabrication techniques and procedures that are intrinsically three-dimensional.

3D microstructures are expected to be made most efficiently using 3D methods. One of the methods that is still in its infancy, but already shows a great potential, is laser lithography. Rather than recording sequences of 2D patterns, this technique allows structuring of both surfaces and bulk materials, and can directly write 3D patterns with fine resolution. The structuring occurs via exposure of material to the powerful optical field of a focused laser beam, or to the periodic interference pattern of several laser beams. The optical field photomodifies material via a wide range of optical mechanisms, from optical damage of transparent dielectrics to optically triggered, chemically amplified cross-linking of polymers in photoresists or photosensitive liquid resins. The principles and physical mechanisms allowing 3D laser lithography of photoresists and resins is the main focus of this review. Demonstration of various micro- and nanostructures in this work is mainly aimed at emphasizing the capabilities of the fabrication techniques, methods, and materials. Functionality of these structures is mentioned only briefly; more details regarding the latter can be found in the cited references.

2
The Techniques of Laser Lithography

Practical laser processing of materials involves a batch of procedures, each of which has its own requirements: preparation of the initial material (with parameters customized for the intended fabrication method), its proper optical exposure, and post-processing aimed at developing or refining the exposed material (e.g., thermal annealing, chemical development, or etching).

The field of laser lithography (also often called laser microfabrication) branched out from multiphoton excitation microscopy and has become established during the last decade. The principles of two-photon microscopy, which has enabled high-resolution 3D imaging [3, 4] and optical memory [5], were gradually adapted for 3D laser lithography [6]. A relevant collection of seminal papers on the field can be found in [7].

Nonlinear absorption is the key mechanism responsible for 3D structuring of materials, including photoresists and photosensitive resins. Optical nonlinearities take place when intensity of the irradiating electrical approaches that of molecular coupling, which occurs at the levels of approximately 10^{10} V/m or ~ 100 GW/cm^2. Detailed description of optical properties of polymers can be found in the literature [8]. Among the optical nonlinearities, multiphoton absorption processes are the most common, in particular, the two-photon absorption (TPA), predicted by Göppert-Mayer in 1931 [9]. These processes are the workhorse of laser lithography. More complicated optical response is also often encountered, for example field enhancement and nonlinear absorption due to collective oscillations of the optically induced dipoles, which is an inherently thresholdless process occurring at considerably lower irradiance levels [8].

At the present time there exist two distinctly different kinds of laser lithography. Direct laser writing (DLW) is a sequential method in which structures are drawn by translating the focal spot of a tightly focused laser beam. Recording of multiple-beam interference patterns, also called holographic lithography, is a parallel method in which the entire structure is exposed simultaneously. In both methods photomodification patterns can have arbitrary dimensionality and high spatial resolution on the subwavelength level. They also allow tuning the processing wavelength to achieve a required excitation mechanism, from single-photon absorption for surface processing to two- or multiphoton absorption for volume processing [10]. Hence, the light sources and the processed materials must match each other's spectral window. Unlike 2D structures, 3D structures cannot be replicated by molding, and fast fabrication throughput at low cost is of primary importance.

Due to these reasons, accurate control over the light delivery into the material and exploitation of the linear and nonlinear optical processes are crucial for the successful implementation of laser lithography. Here we survey the methods and principles allowing control of the optical fields required for laser lithography, including light delivery, optical field localization, its intensity enhancement, and quantum optical effects. The main aim of this review is to show the usefulness of laser lithography in terms of its versatility and high resolution.

2.1
Direct Laser Writing

The principle of direct laser writing is explained in Fig. 1. Optical exposure and nonlinear absorption is induced in the focal region of a tightly focused laser beam, whose size is defined by the point-spread function (PSF) of the nonlinear process and may be significantly smaller than the wavelength of the recording laser. The focal region can be smoothly translated in the bulk of photosensitive material, where linear features having the desired shape can be recorded. After the exposure, the development procedure removes the unexposed parts of material, leaving solid features with subwavelength size. The setup of 3D laser lithography used in our laboratories employs a pulsed output of a Hurricane X system (Spectra-Physics) with $\tau_{pulse} = 120\,fs$ and $\lambda_{pulse} = 800\,nm$ as the light source. The fabrication is performed in an optical microscope (Olympus IX71) equipped with oil-immersion objective lenses

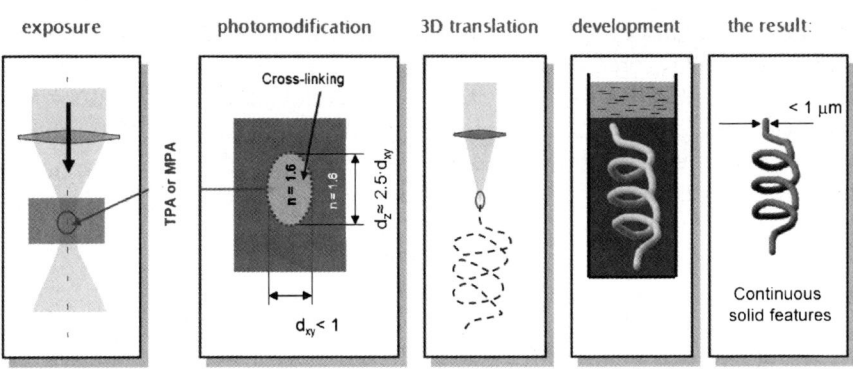

Fig. 1 Principle and implementation of DLW in a positive photoresist or liquid resin (from *left* to *right*)

(magnification $60\times$ and $100\times$, numerical apertures $NA = 1.4$ and 1.35, respectively). 3D drawing is accomplished by mounting the samples on a piezoelectric transducer (PZT) controlled 3D translation stage (Physik Instrumente PZ48E) which has a maximum positioning range of up to $50\,\mu m$ and an accuracy of several nanometers. The samples are films of SU-8 (Microchem), spin-coated to a $50\,\mu m$ thickness on glass substrates. Single-photon absorption in SU-8 is negligible at λ_{pulse}, but becomes dominant at $\lambda = 400\,nm$ [12]. Therefore, two-photon absorption is responsible for the photomodification by the cross-linking of a polymer, which renders SU-8 stable against subsequent chemical development. The development leaves solid exposed parts, which have a refractive index of $n \approx 1.6$ (at infrared wavelengths) suspended in air ($n = 1$). Therefore periodic structures recorded by DLW in SU-8 can be useful in photonics.

In terms of beam delivery, the DLW method is based on optical microscopy, confocal microscopy [4, 6, 13] and laser tweezers [14] (for reviews on laser tweezers see [15, 16]). These techniques allow for a high spatial 3D resolution of a tightly focused laser beam with optical exposure of micrometric-sized volumes via linear and nonlinear absorption. In addition, mechanical and thermal forces can be exerted upon objects as small as $10\,nm$; molecular dipolar alignment can be controlled by polarization of light in volumes of with submicrometric cross-sections. This circumstance widens the field of applications for laser nano- and microfabrication in liquid and solid materials [17–22].

Figure 2 shows a scanning electron microscopy (SEM) image of a complicated 3D structure recorded by DLW using exposure by a sequence of $0.5\,nJ$ pulses separated by $100\,nm$ from each other. The structure has a so-called circular spiral geometry, highly regarded as a building block of PhC structures with strongly pronounced PBG properties. This topology is derived from the diamond structure and has a spectrally wide $\sim 28\%$ (width-to-center) bandgap, if fabricated in a high refractive index material of $n = 3.5$ (e.g., sil-

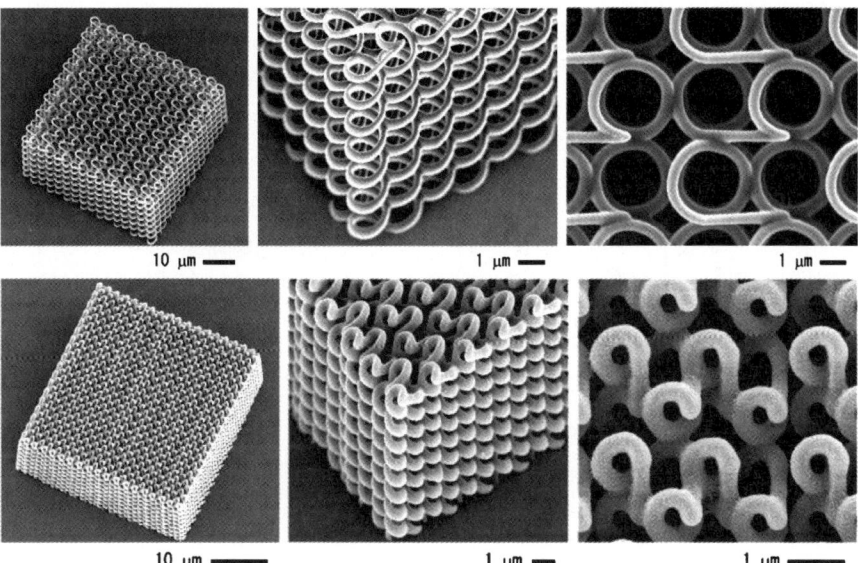

Fig. 2 SEM images of non-interlaced ($a \simeq L$) spiral structures recorded by direct laser writing in SU-8 [11] and their transmission/reflection spectra. *Upper row*: spirals with lateral period $a = 3.0\,\mu m$, arm length $L = 2.7\,\mu m$, vertical spiral pitch $c = 4.32\,\mu m$, structure volume $16 \times 16 \times 8$ spiral periods. (*Lower row*: $a = 0.95\,\mu m$, $L = 0.7\,\mu m$, $c = 1.484\,\mu m$, $36 \times 36 \times 12$

icon) [23]. The high structural quality of the sample shown in Fig. 2 can be inferred from photonic stop-gaps (PSG), recognizable in the optical transmission and absorption spectra of the samples shown in Fig. 3.

These structures were recorded by a vectorial focal spot scanning in a spiral-by-spiral method rather in a raster layer-by-layer mode using a PZT stage. Such spiral structures fabricated in SU-8 have optical spot bands in near-IR [24], telecommunication [25], and 2–5 μm-IR region [26] or can be used as templates for Si infiltration [11]. It is obvious, that direct laser scanning is well suited for defect introduction into 3D PhC, as demonstrated in resin where a missing rod of a logpile structure resulted in the appearance of a cavity mode in an optical transmission spectrum [27].

2.2
Holographic Lithography

The optical scheme of holographic lithography is schematically illustrated in Fig. 4. The beam of the same laser system as used for DLW recording passes through a diffractive optical element (DOE), where it is split into a set of diverging multiple beamlets by diffraction. The set of beamlets is subsequently collimated by a lens, and passed through the transmission mask, which selects

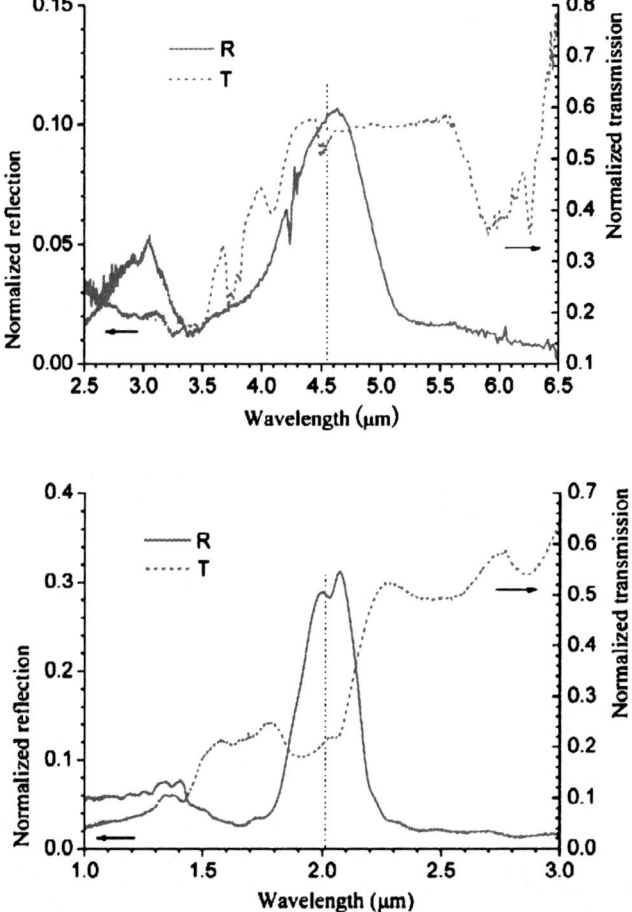

Fig. 3 Infrared optical transmission and reflection spectra of the samples shown in Fig. 2; the spectrally matching transmission dips and reflection peaks represent photonic stop-gaps. *Upper* and *lower plots* correspond to *upper* and *lower rows* in Fig. 2

the beamlets required for obtaining the desired 3D interference pattern. As the beamlets propagate in free space, they can be optionally passed through variable phase retarder (PR) plates, which set their mutual phases. Phase control can be used to alter the 3D interference patterns. The parallel, phase-controlled set of beamlets is then focused into the photoresist sample using a focusing lens having numerical aperture lower than in the case of DLW. Typically, a lens with NA = 0.75 is used. Multiple beams formed by DOE have an advantage of larger mutual spatial overlap compared to that of the beams obtained from a beamsplitter. The DOE produces multiple, mutually coherent laser beams from a single input beam. The beams arrive to the focal spot of the lens as nearly plane waves with vanishing optical path difference. This is especially

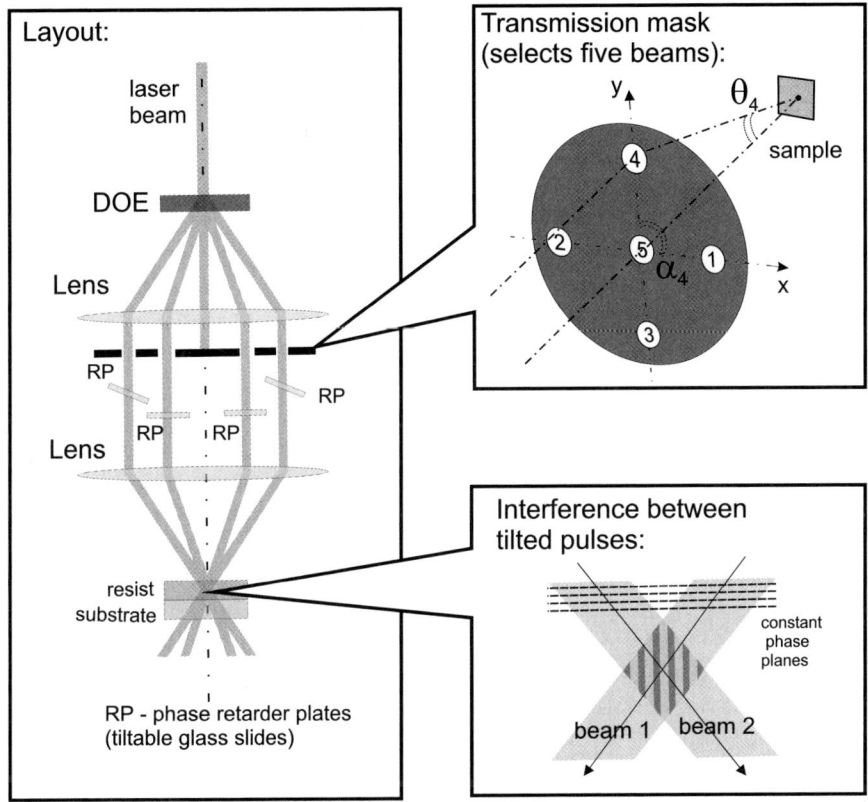

Fig. 4 Schematic layout of the optical setup for holographic lithography using diffractive DOE, and phase control of the interfering beams. The callouts illustrate the selection of the beams using transmission mask, their incidence directions (*top*), and extended transverse spatial range of the interference region using tilted pulses produced by DOE (*bottom*)

important when ultrashort laser pulses are used [28]. Overlap of two coherent ultrashort pulses is determined by their longitudinal dimension, $l_p = c\tau_p$, which is about 30 μm for a 100 fs duration pulse. Hence, the width of the spatial overlap is $w = c\tau_p/\sin(\theta/2)$ for a two-beam interference. For diffracted pulses, their constant intensity surface does not coincide with the the wave front, resulting in tilted pulses, whose spatial overlap is given by $w = d/\cos(\theta/2)$, where d is the diameter of the beams. Thus, mutual coherence and zero time-delay are maintained across the entire cross-sectional areas of the tilted pulses despite their different propagation directions, and interference occurs in the entire focal spot. Large patterns can be recorded even with tight-focusing optics having a numerical aperture of $NA > 0.5$.

Quality of the DOE is crucial for the recording of high-fidelity patterns, especially when phase control is involved. It can be evaluated using shear-interferometry by combining the front- and back-reflected beams on a CCD camera. The uniform interference fringes signify the uniform phase (thickness) of the DOE, as is illustrated in Fig. 5.

Spatial intensity distribution of the interference field depends on the phase difference between the beams $\delta_{ij} = k_i - k_j$ (in vector notation) and their mutual angles $\theta_{i,j}$ (where i, j are the beam index) as:

$$I = \sum_j E_{0j}^2 + \sum_{i>j} 2E_{0i}E_{0j}\cos(\theta_{i,j})\cos(\delta_{ij}). \tag{1}$$

Here E_0 is the amplitude of electrical field of the beam. Equation 1 defines spatial intensity distribution in the interaction region as a function of beam phase and incidence angle. For practical applications, the contrast of the intensity distribution $I(x, y, z)$ should be maximized. This can be achieved by optimizing the intensities of the beams, E_{0i}^2, and their phases. The orientation of linear polarizations of the beams are also important and would enter Eq. 1 as additional $\cos(e_i \cdot e_j)$ factors for the polarizations $e_{i,j}$ (of beams i and j, respectively), becoming more significant at larger incidence angles. For s-polarization perpendicular to the plane of incidence, the highest contrast ($\cos(e_i \cdot e_j) = 1$) is reached. However, complex multibeam interference patterns are usually formed by both s- and p-polarizations, and only their relevant components should be considered. It is noteworthy, that chiral patterns can be created by interference of circularly and linearly polarized beams.

The phase control makes it possible to modify and translate the interference pattern with a fine subperiod resolution [29, 30]. Also, very complex patterns can be formed by combining linearly and circularly polarized beams [31–33] and by using several exposures of the same sample by different patterns. Equation 1 can be recast in the form more convenient for theoretical simulation and visualization of the interference patterns of multiple beams with adjustable phases:

$$I(r) = \sum_{n,m} E_n e^{-i(k_n \cdot r + \delta_n)} \cdot E_m^* e^{i(k_m \cdot r + \delta_m)}, \tag{2}$$

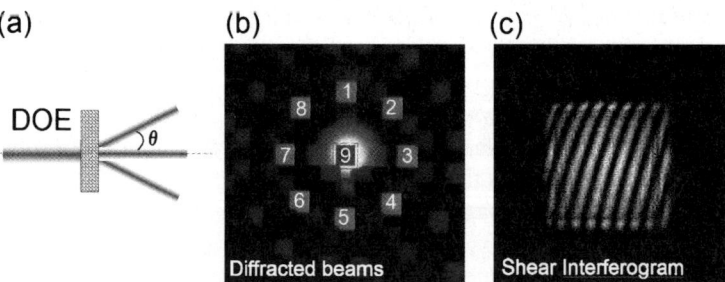

Fig. 5 a Diffraction in a DOE. **b** Nine-beam diffraction pattern obtained by laser illumination of a DOE made of silica ($\theta = 3.82°$). **c** Shear interferometry image made by interference from the front and back side of the DOE shown in (**b**)

where E is the E-field vector (* designates a complex conjugate), r is the radius-vector, k the wave vector, and $n = m$ represents the number of interfering beams.

Phase control provides an additional avenue for fine tuning of the interference patterns [34–36]. Interference of several coherent beams can be simulated numerically by summation of plane waves according to Eqs. 1 and 2. This is a helpful method for designing structures with desired architecture.

In practice, phases of the beams are adjustable by tilt angle, φ, of phase retarder plates (glass plates) inserted into the beamlets. The tilt defines the the optical path $\Delta x = nd/\cos(\varphi)$; typically cover-glass slides with refractive index of $n = 1.5$ and thickness of $d = 180\,\mu m$ suffice as variable phase-retarders.

Figure 6 shows SEM images of the structure recorded in photoresist SU-8 using interference of five beams with the same phase. The resulting structure has a body-centered tetragonal symmetry. It was demonstrated recently that the same optical layout, with the addition of phase control, allows recording of so-called woodpile structures that are highly promising as PBG materials [29, 37]. Such structures have open architecture with good permeability to the liquid developer, resulting in high quality samples suitable as templates of the structures for photonic and microfluidic applications.

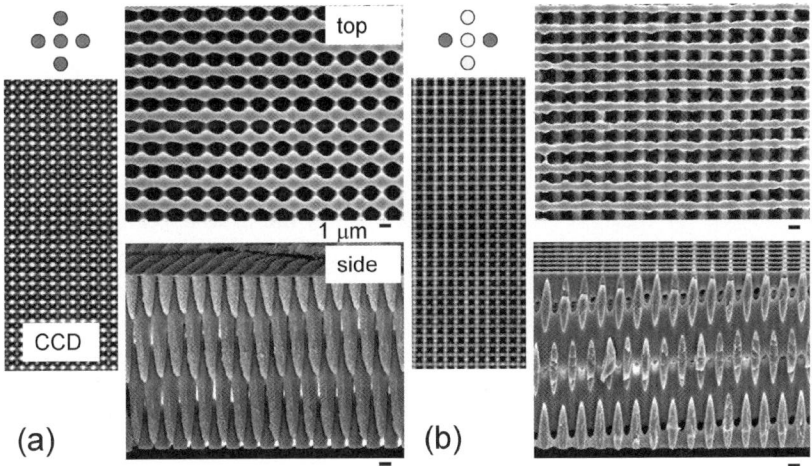

Fig. 6 Structures recorded by five-beam interference in SU-8. The interference images were recorded on an auxiliary CCD to observe the pattern [29]. Top and side-view SEM images of structures **a** without phase control of interfering beams, and **b** with the phase difference of $\pi/2$ between three central beams and the two side beams. The numerical aperture of the lens was 0.75, the exposure time was 30 min at a 35 mW total irradiation power (all beams, at a 1 kHz repetition rate). The pulse duration was 180 fs, the central wavelength was 800 nm

3
Delivery of the Optical Radiation

The light intensity distribution at the focus of the lens is the main factor defining resolution of DLW, especially because optical nonlinearities are involved in the photomodification. Light intensity in the focal region depends on the numerical aperture of the lens, duration of the laser pulse, spatial mode of the laser beam, aberrations, and propagation effects. Though the smallest focal spot achieved by an optical system is about half of the diffraction-limited spot size [38], spatial resolution several times higher was achieved using laser lithography due to the nonlinearities and the existence of threshold intensity for photomodification.

3.1
Focal Spot Size of a Gaussian Beam

The Gaussian beam approximation is usually adequate for description of pulsed and continuous-wave (cw) laser pulses and beams not clipped by apertures of optical elements. However, when the laser beam is clipped, for example by the entrance pupil of objective lens, the central part of the clipped Gaussian beam can often be used successfully for laser lithography, because ellipticity of the focal spot (the ratio between the long and short axes of the ellipsoidal focal region) is minimized for clipped beams. A Gaussian beam has the smallest possible angular diameter caused by diffraction $2\lambda/d$, where the diameter $d = \pi w_0$ is determined by the waist of the beam, w_0. The diameter of a truncated Gaussian beam approaches that of a plane wave diffracted on the aperture having diameter $2.44\lambda/d$ [39].

For a well-controlled 3D exposure, it is necessary to control the light intensity distribution at the focus, which is described by the point-spread function (PSF). Even for tight, aberration-free focusing, the axial spreading of PSF is larger than the lateral spreading [40]:

$$l = d\frac{\sqrt{3 - 2\cos\alpha - \cos 2\alpha}}{1 - \cos\alpha}, \qquad (3)$$

where the lateral diameter of the focal spot size is defined by the Airy disk for a plane wave focusing, $d = 1.22\lambda/NA$. The numerical aperture $NA = n\sin\alpha$, where n is refractive index at the focus, λ is the vacuum wavelength, and α is the half-angle of the focusing cone. Equation 3 is valid for focusing by large numerical aperture ($NA > 0.7$) optics.

Axial elongation of the PSF may also result from the pulse tilt acquired after their passing through prisms and gratings, for example, in pulse compressors used in the standard chirped pulse amplification (CPA) scheme [41]. The influence of spherical aberrations on the PSF is discussed in Sect. 3.1.2.

It is instructive to identify regions of the focal spot where intensity is higher than the threshold of the photomodification process concerned. As mentioned earlier, the photomodification may involve polymerization, crystalline phase change, melting, structural modification, defect formation, etc. Though the final dimensions of the photomodified region depend on the material response, they follow closely the spatial intensity profile, especially near the photomodification threshold. The intensity envelope of a Gaussian beam is:

$$I(r, z) = I_0 \frac{w_0^2}{w(z)^2} e^{-2\left(\frac{r}{w(z)}\right)^2}, \tag{4}$$

where waist (radius) of the beam w_0 determines its lateral cross-section along the z-axis, and is given by:

$$w(z) = w_0 \sqrt{1 + \left(\frac{z\lambda}{w_0^2 n\pi}\right)^2} \equiv \sqrt{\frac{\lambda}{n\pi} \left(z_R + \frac{z^2}{z_R}\right)}, \tag{5}$$

where n is the refractive index at the focus, λ is the wavelength in vacuum, and:

$$z_R = \frac{n\pi w_0^2}{\lambda}, \tag{6}$$

is the Rayleigh length at which waist increases by the factor of $\sqrt{2}$.

It is usually instructive to compare lateral and axial dimensions of the photomodified region and the focal spot. The size of the focal region can be evaluated from Eq. 2. Along the lateral direction, the intensity reaches the level of I_0/e^2 at the position characterized by $r = w_0$. Along the axial direction, the same intensity is reached at the position z_R. Hence, it is convenient to define the lateral spot size (diameter) at the $I_0/2$, or full-width at half maximum (FWHM) level. This size is smaller than that defined at the I_0/e^2-level by factor of $\sqrt{\ln(2)/2} \simeq 0.59$.

By setting $I(r, z) = I_{th}$ in Eq. 4, one can find the diameter, D, and the length, L, of the focal region where intensity exceeds the threshold value:

$$D(r) = w_0 \sqrt{2 \ln\left(\frac{I(r)}{I_{th}}\right)}, \quad L(z) = 2z_R \sqrt{\frac{I(z)}{I_{th}} - 1}. \tag{7}$$

Expressions accounting for the nonlinear absorption, which is the main mechanism enabling 3D laser lithography, can be easily obtained by assuming the photomodification to be dependent on the intensity as $\propto I^N$:

$$D(r) = w_0 \sqrt{2 \frac{1}{N} \ln\left(\frac{I(r)}{I_{th}}\right)}, \quad L(z) = 2z_R \sqrt{\left(\frac{I(z)}{I_{th}}\right)^{\frac{1}{N}} - 1}, \tag{8}$$

where N is the order of nonlinear absorption. Assuming paraxial focusing conditions where optical rays converge at low angles with respect to the optical axis, $w_0 = \lambda/(\pi NA)$. Here, $NA = 1/2f_\#$ is the numerical aperture, $f_\# = f/a$ is the f-number, f is the focal distance, and a is the diameter of the beam. Beam truncation by the finite aperture of the focusing optics is not considered here. Implications of Eqs. 7 and 8 can be demonstrated by the following numerical example. For a lens with $NA = 1.35$ and at a peak intensity in the focal spot exceeding the threshold value by approximately 5%, lateral and axial diameters of the focal region shrink to about 30 and 90 nm, respectively. Thus, careful intensity control in the presence of TPA or other nonlinear absorption process allows one achieve spatial resolution even higher than the earlier confirmed benchmark value of about 100 nm [42]. However, due to the nonlinearity, the opportunities for practical 3D lithography with such resolution depend strongly on the stability of the laser.

Although paraxial approximation becomes unsuitable for higher-NA optics and for non-Gaussian beams, the above insights should remain qualitatively valid in these cases as well. Since only the above-the-threshold intensity part of the spatio-temporal envelope of the beam is important for photomodification, usually this part can be reasonably well approximated by a Gaussian.

3.1.1
Resolution Limit for Plane-Wave Focusing

Figure 7 shows an aberration-free intensity distribution at the focus of a typical objective lens similar to that used for DLW lithography. Calculations were carried out using a vectorial Debye theory, which accounts for the polarization effects. For the linearly polarized wave it can be seen that the spot is elongated along the polarization vector. To reduce this asymmetry, a $\lambda/4$-plate can be used to convert the polarization of the incident beam to circular, which can be interpreted as a combination of two mutually perpendicular linearly polarized components. Thus, width of the photomodified line becomes independent of the beam scanning direction in the sample.

3.1.2
The Role of Spherical Aberrations

In reality, the idealized intensity distribution in the focal region, described in the preceding sections, is affected by distortions, which in turn affect resolution of laser lithography. For laser beams centered on, and propagating along, the optical axis of an optical system consisting of lenses, spherical aberrations are the most common distortions. Spherical aberrations evolve due to the refractive index mismatch along the propagation direction at the entrance face of the sample, and become stronger with increasing focusing depth d.

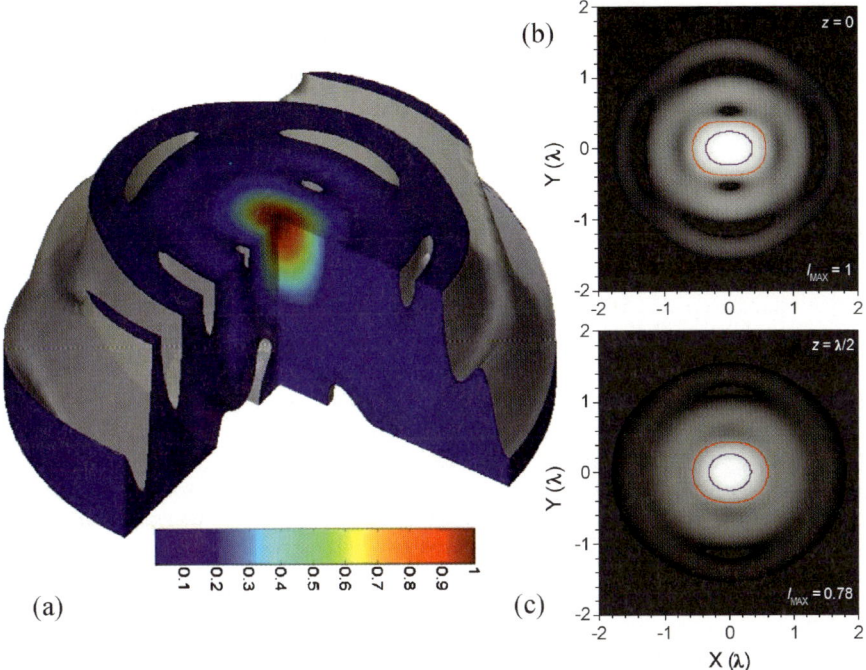

Fig. 7 Plane wave focusing by a $NA = 1.35$ objective lens, calculated using vectorial Debye theory. **a** The normalized 3D intensity distribution with the cutoff threshold at 1% intensity. The lateral cross-sections are plotted on a log-scale at the axial positions $z = 0$ (**b**) and $z = \pm\lambda/2$ (**c**), respectively. Contour lines are plotted at 0.5 (*inner*) and $1/e$ (*outer*) levels, respectively. Polarization of the plane wave was horizontal (along x)

The point spread function (PSF) resulting under these circumstances was derived in using scalar Debye theory [43]:

$$\Phi(\theta_1, \theta_2, d) = - kd \left(n_1 \cos(\theta_1) - n_2 \cos(\theta_2) \right) , \tag{9}$$

where $n_{1,2}$ and $\theta_{1,2}$ are the refractive indexes and angles of the side rays along the light propagation, $k = 2\pi/\lambda$ is the wave-vector in vacuum. PSF in cylindrical coordinate system is given by:

$$I_s(r_2, z_2, \theta_1, \theta_2, d) = \left| \int_0^\alpha P(\theta_1) \sin(\theta_1) \left(t_s + t_p \cos(\theta_2) \right) J_0 \left(k_1 r_1 n_1 \sin(\theta_1) \right) \right.$$
$$\left. \times \exp \left(- i\Phi(\theta_1, \theta_2, d) - i k_2 z_2 n_2 \cos(\theta_2) \right) d\theta_1 \right|^2 , \tag{10}$$

where $t_s(\theta_1) = 2 \sin(\theta_2) \cos(\theta_1)/\sin(\theta_1 + \theta_2)$ and $t_p = 2 \sin(\theta_2) \cos(\theta_1)/(\sin(\theta_1 + \theta_2) \cos(\theta_1 - \theta_2))$ are the Fresnel coefficients for s- and p-polarizations, respectively; $NA = n_1 \sin(\alpha)$ is the numerical aperture of objective lens, α is the half-angle of the focusing cone, $P(\theta_1) = \sqrt{\cos(\theta_1)}$ is the apodization function

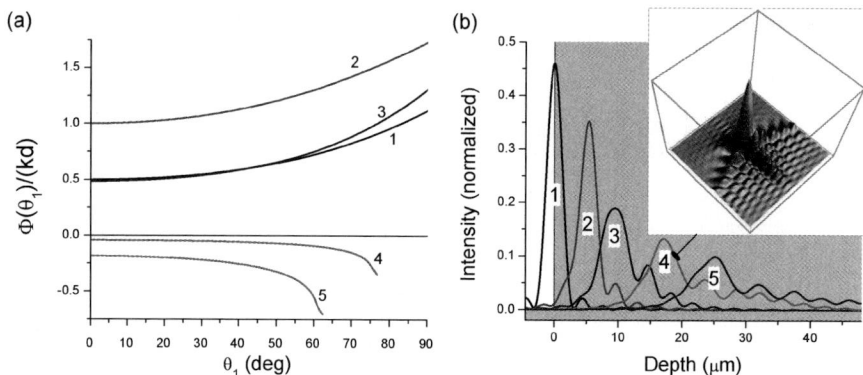

Fig. 8 **a** Normalized aberration function vs. different incidence angles, θ_1, for focusing from refractive index n_1 into n_2: air-to-glass/polymer $n_1 = 1$ to $n_2 = 1.5$ (*1*), $n_1 = 1$ to $n_2 = 2$ (*2*), $n_1 = 1.515$ (immersion oil) to $n_2 = 2$ (*3*), immersion oil-to-silica/polymer $n_1 = 1.515$ to $n_2 = 1.47$ (*4*), and immersion oil-to-water $n_1 = 1.515$ to $n_2 = 1.33$ (*5*). Truncation of curves (*4, 5*) is due to the total internal reflection. **b** Axial intensity distribution caused by spherical aberration when focused from immersion oil $n = 1.515$ into $n_2 = 2$ dielectric medium (corresponds to curve (*3*) in **a**) aiming at different depths: 0 (*1*) (aberration-free), 5 (*2*), 10 (*3*), 20 (*4*) (*inset* shows 3D intensity distribution), and 30 μm (*5*). The wavelength $\lambda = 0.8$ μm, $NA = 1.35$ ($\theta_1 \simeq 63°$), and apodization function of objective lens obey the sine condition, i.e., $P(\theta_1) = \sqrt{\cos(\theta_1)}$

obeying sine conditions (the commercial objective lenses are designed to satisfy the sine condition), and $J_0(...)$ is the 0th-order Bessel function of the first kind.

Figure 8a and b illustrate Eqs. 9 and 10, respectively. The normalized aberration function $\Phi/(kd)$ is plotted for several realistic refractive index contrasts in Fig. 8a. The sign of the aberration function changes upon change of the refractive index contrast. The intensity becomes increasingly smeared with focusing depth and refractive index Fig. 8b.

3.2
The Role of Pulse Duration

Duration of subpicosecond laser pulses arriving to the focal spot inside a transparent material is modified due to the pulse chirp and group velocity dispersion (GVD) in various optical elements encountered by the pulse during its propagation. Ultrashort laser pulses can be pre-chirped by a diffractive grating used in the pulse compressor at the output of the amplifier to produce shortest pulse duration at the focus. Such optimization is usually carried out by observation of the dielectric breakdown threshold in solid or gaseous dielectrics, e.g., glass or air [44]. The conditions under which the the breakdown occurs at the smallest energy of the incident pulse correspond to the shortest pulse at the focus.

Pulse duration can be measured reliably using detection of both amplitude and phase of its spectral components. This can be accomplished using frequency-resolved optical gating (FROG) or interferometric auto-correlation techniques. A versatile and simple FROG technique is available commercially as a grating-eliminated no-nonsense observation of ultrafast incident laser light e-fields (GRENOUILLE) [45].

3.3
Nonlinear Effects

3.3.1
Self-focusing

Self-focusing is not usually desired for laser lithography and should be avoided by maintaining the power of cw or pulsed laser beam below the self-focusing threshold in the material. Its value depends on the nonlinear part of the refractive index, n_2, $(n = n_0 + n_2 I)$, as follows [46]:

$$P_{cr} = \frac{\lambda^2}{2\pi n_0 n_2}. \tag{11}$$

Self-focusing occurs when power, P_0, exceeds the critical self-focusing power P_{cr}. A Gaussian beam experiences self-focusing after propagating distance, L_{sf} [46]:

$$L_{sf} = \frac{2\pi n_0 w_0^2}{\lambda} \left(\frac{P_0}{P_{cr}} - 1 \right)^{-1/2}, \tag{12}$$

where w_0 is the waist (radius) of a Gaussian beam. For example, in fused silica the self-focusing threshold is 3 MW at $\lambda = 1$ μm ($n_0 = 1.45$; $n_2 = 3.54 \times 10^{-16}$ cm^2 W^{-1}). Assuming $P_0 = 2P_{cr}$ and $w_0 \sim \lambda$, one can estimate the self-focusing distance to be 9λ. Therefore, one can obtain intensities above 10^{13} W cm^{-2} in the focal plane, and simultaneously prevent the self-focusing by using high-NA optics. Typical n_2 values in resins and photoresists are larger by ten and more times. However high-fidelity 3D laser lithography is still possible using tight focusing with $NA > 0.8$ optics. On the other hand, focusing by a low NA lens may be used to induce self-focusing intentionally for self-guided recording of light filaments, which can be subsequently used as waveguides [47].

3.3.2
Nonlinear Absorption

Among the mechanisms of nonlinear absorption, TPA is the most important and widely used for 3D laser lithography in resins and resins. TPA requires smaller incident powers than higher-order processes, for example multiphoton absorption (MPA), or electronic excitation via tunneling and avalanche. Utilization of nonlinearities other than TPA will be discussed later.

Linear and nonlinear absorption at the frequency, ω ($\omega = 2\pi\lambda/c$, where λ is the wavelength and c is speed of light) can be described by:

$$\frac{dI}{dz} = -\sigma_{ab}(\omega)NI - \sigma_{2\gamma}(\omega)NI^2 , \tag{13}$$

where I is the intensity, N is the number density of absorbers in the medium (e.g., atoms or molecules), z is direction of light propagation, σ_{ab} is the linear absorption cross-section (in SI units of m^2), and $\sigma_{2\gamma}$ is the TPA cross-section (m^2 W^{-1}). The solution of Eq. 13 with initial condition $I(0) = I_0$ is [48]:

$$I(z) = \frac{N\sigma_{ab}(\omega)}{-\sigma_{2\gamma}(\omega)N + \dfrac{N\sigma_{ab}(\omega) + \sigma_{2\gamma}(\omega)NI_0}{I_0}\exp[N\sigma_{ab}(\omega)z]} , \tag{14}$$

which becomes the familiar Beer–Lambert law $I(z) = I_0\exp[-\sigma_{ab}(\omega)Nz]$ for linear absorption. Neglecting the linear absorption yields $I(z) = I_0/(1 + \sigma_{2\gamma}(\omega)NI_0z)$ for TPA. The nonlinear absorption coefficients can be expressed via the corresponding absorption cross-sections:

$$\alpha_{2\gamma}\ [\text{cm/W}] = \frac{\sigma_2\ [\text{cm}^4\text{s}]N_A d_0 \times 10^{-3}}{\hbar\omega} , \tag{15}$$

where N_A is the Avogadro number, d_0 is the molar concentration (mol L^{-1}), and $\hbar\omega$ the photon energy.

Synthesis of materials with high TPA efficiency is being actively pursued for the applications such as photosensitizers, fluorescence and second-harmonic markers in 3D molecular spectroscopy, and photo-initiators of two-photon polymerization in photoresists and liquid resins. For example, a hydrophobic dye $C_{40}H_{54}N_2O_2$, known under acronym MBAPB, with molar weight of 580.4155 g M^{-1} [49, 50], is one of the most efficient two-photon absorbers suitable for inducing negative tone photopolymerization. Its absorption, emission and TPA spectra measured by Z-scan techniques are shown in Fig. 9. The 700–800 nm spectral window is well suited for the widely available Ti:sapphire lasers. The TPA cross-section in this spectral range reaches several hundreds of Göppert-Mayer GM units (1 GM = 10^{-50} cm^4 s). Efficient two-photon absorbers for initiation of positive tone [51] and inorganic–organic [52] photopolymerization have been also developed.

It is noteworthy that nonlinearity of the absorption, required for 3D microfabrication can be provided via thermal mechanisms [53, 54]. In the case of tightly focused laser pulses, linear absorption is most efficient at the focus, where local heating can create the conditions required for polymerization. Usually the absorption increases with temperature and thermal polymerization may become dominant at the focus. It is usually difficult to confirm the TPA mechanism from the direct transmission measurements due to the nar-

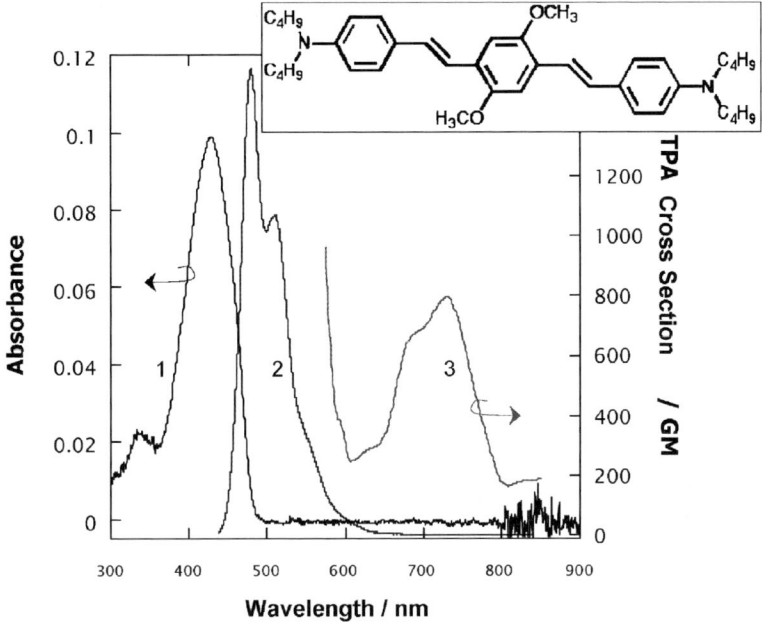

Fig. 9 Absorption (*1*), emission (*2*), and TPA (*3*) spectra of MBAPB in toluene (courtesy of Dr. K. Kamada). Inset shows the chemical structure of MBAPB

row dynamic range between the onset of polymerization and the dielectric breakdown of resin or resist, and bubble formation [55]. The departure of transmission from the linear law signifies presence of TPA [56]. Moreover, photopolymerization, which is exothermic reaction, changes conditions for the polymerization at the focus, especially when the focal spot size is comparable to the wavelength. Ionization of the focal volume and presence of high density electrons is expected to alter the usual pathways of TPA-initiated photopolymerization [57].

The nonlinear response of material to optical excitation is the main requirement for formation of 3D patterns by photopolymerization. The application potential of such patterns in microphotonic, photonic crystal, MEMS, and microfluidic applications is described in our contributions to [58].

4
Field Enhancement

The TPA efficiency depends on the intensity of the optical field. Besides the adjustment of the source intensity, the intensity can be also varied by adjusting the degree of field localization in space and time. Spatial localization of the field can be achieved most easily by focusing (Sect. 3.1). Here we consider methods other than focusing that allow enhancement of the optical field, and the absorption processes induced by it, without the increase of the incident power. These mechanisms are near-field enhancement by metallic nanostructures, particles and apertures, and enhancement by spatial ordering of the absorbing molecules.

4.1
Fresnel Enhancement

A simple and practical way to achieve the field enhancement is to use backside illumination of a dielectric plate, for instance a cover glass, in a standard DLW geometry with an oil-immersion focusing lens. According to the Fresnel formulas for the right angle incidence ($\theta_i = 0°$), the coefficients of the in-plane ($\parallel$) polarized amplitudes of transmitted and reflected electric fields are, respectively:

$$t_\parallel = \frac{2n_i}{n_i + n_t} \tag{16}$$

$$r_\parallel \equiv 1 - t_\parallel = \frac{n_t - n_i}{n_i + n_t}, \tag{17}$$

where the $n_{i,t}$ are the refractive indices of the material on the incidence and transmission sides, respectively. The field component perpendicular to the plane of incidence is $t_\perp = t_\parallel$ and $r_\perp =- r_\parallel$ (the corresponding intensities are r^2 and t^2). The enhancement occurs when light passes from the medium with a larger refractive index to the medium with lower refractive index. In the case of the glass–air boundary ($n_i = 1.5$, $n_t = 1$) the transmitted intensity is enhanced by a factor of $t^2 = (3/2.5)^2$; while for the air–glass boundary loss occurs in transmission $t^2 = (2/2.5)^2$. The physical reason for this enhancement is the phase shift at the boundary, depending on the refractive index change, while the energy is conserved: $t^2 + r^2 = 1$. Such enhancement can even cause ablation at the back-side of glass plates due to tensile stress (in glass the tensile strength is only $\sim$30–60 MPa and depends on the microcrack density, while the compressive strength is comparable to the Young modulus of 70 GPa).

4.2
Dipolar Ordering

The orientational average of dipoles oriented at an angle θ with respect to the direction of the linear light polarization depends on the factor:

$$\gamma_I \equiv \langle \cos(\theta)^4 \rangle = \frac{1}{4\pi} \int_0^{2\pi} \int_0^\pi \cos(\theta)^4 \sin(\theta)\, d\theta\, d\varphi, \tag{18}$$

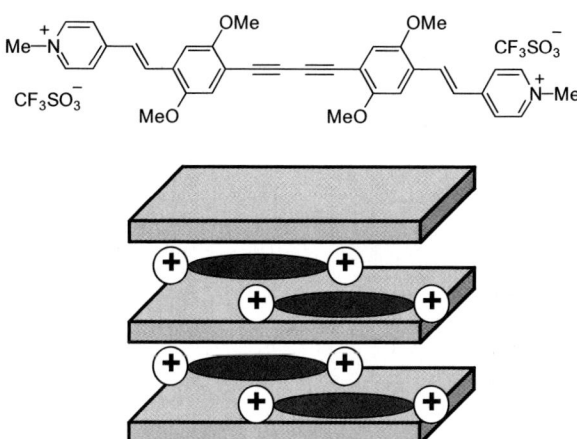

Fig. 10 Structure of cationic MPPBT molecule and schematics of the 2D confinement between clay lamellas (courtesy of Dr. K. Kamada)

where θ, φ are spherical coordinates. Perfectly parallel dipoles produce strongest absorption according to $\gamma_I = 1$, which can be considered as a one-dimensional (1D) ordering. On a 2D plane of randomly oriented dipoles, the average is $\gamma_I = 3/8$, and becomes even smaller for the 3D case ($\gamma_I = 1/5$). Figure 10 illustrates schematically the 2D molecular intercalation of MPPBT molecules (see [59] for more details) inside a clay sample of lamella structure with nanometric separation between the sheets. TPA properties of this system will be reported in the future (Kamada, unpublished results).

4.3
Near-Field Effects

Ordered 3D structures consisting of small features with sizes comparable to the optical wavelengths cause diffraction of the incident beams or pulses. The near-field pattern resulting from the multiple interference can exhibit regions of local field enhancement that may be used for inducing photopolymerization or other kinds of photomodification. Figure 11 shows calculated intensity pattern resulting from diffraction of a plane on an array of empty channels inside material with refractive index $n = 1.5$ Such channels can be also recorded using laser. The diffracted pattern clearly shows that the field amplitude is enhanced by several times.

4.4
Plasmonic Effects

Huge local enhancement of the near-field intensity by factors of $\sim 10^5$–10^6 occur at sharp tips and edges of nanoparticles of noble metals and on rough

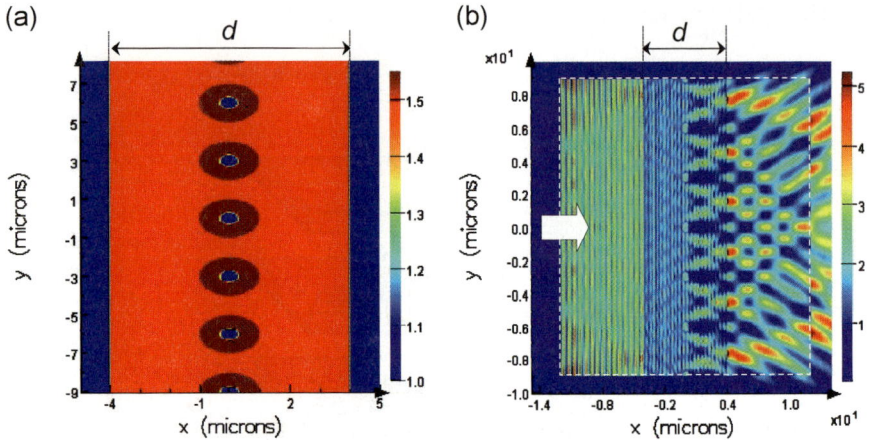

Fig. 11 2D model structure of empty channels inside a material with refractive index, $n = 1.5$. The refractive index distribution (**a**) and the corresponding near-field pattern of light intensity E^2 distribution was calculated by the FDTD technique. The *arrow* marks the direction of the plane wave $E(0, 1, 0)$ incidence, d is thickness of the sample. The *square* in **b** marks the region of the TFSF source used in the calculations

surfaces due to the localized surface plasmons (LSP). Such enhancement occurs due to the localization of the optical near-field associated with LSP. High field strength enables one to induce TPA and other optical nonlinearities in ultrasmall regions using incident radiation with very low average power.

The ionization of focal volume, or formation of plasma, is expected to alter the usual photochemical material modification pathway, as has been recently demonstrated in photopolymerization of SU-8 resist by femtosecond pulses [57]. In addition, nanometric-sized plasma regions created by the ionization, e.g., at defect sites, have spatio-temporal dynamics of their own. Recently, a model of nanosheet formation from plasma nanospheres in glass has been proposed [60]. Similar conditions are expected in polymers as well. Let us discuss here field enhancement by a metallic nanoparticle (similar arguments are also valid for surfaces containing nanometric features).

For small spherical particles having a radius much smaller than the wavelength of the incident wave ($r \ll \lambda$), a simple electrostatic approach can be used to derive the near-field:

$$E_{\text{loc}}(\omega) = L(\omega)E_0 = \frac{3}{\varepsilon_m(\omega) + 2}E_0 , \tag{19}$$

where $L(\omega)$ is the local field factor, E_0 is the amplitude of the incident light field, and ε_m is the dielectric function of metal. Resonant enhancement of the LSP near-field occurs at frequencies where $\varepsilon_m = -2$. Note that metals have $\varepsilon_m(\omega) < 0$, in contrast to dielectrics where $\varepsilon_d(\omega) \propto 1 - 2 > 0$.

The local field enhancement factor at the resonance is [61]:

$$| L(\omega_{res}) | = \frac{| \varepsilon'_m(\omega_r) |}{\varepsilon''_m(\omega_r)}, \tag{20}$$

where $\varepsilon'_m(\omega)$ and $\varepsilon''_m(\omega)$ are the real and imaginary parts of the dielectric function. In noble metals this factor can reach 10–20 in the vicinity of ω_r, where absorption losses are small $\varepsilon''_m(\omega) \ll 1$.

The field enhancement factor at the surface of the nanosphere is not uniform. The tangential component of field (on the equator of the sphere) is equal to the field inside the sphere multiplied by the factor L (Eq. 19). The normal component (on the poles) is additionally modified by the factor of ε_m, i.e., $L_{poles} = \varepsilon_m \times L$. This enhancement has been suggested as a mechanism relevant to silica breakdown and nanograting formation [60]. It may also explain the anomalous light scattering [62], being maximum along the polarization of the incident light field in the lateral plane (a 90°-rotated distribution would be expected for an oscillating dipole), observed earlier in glass [63].

Figure 12 shows intensity distribution of different electric field components on the plane touching the back-side of the nanosphere. Strong depolarization is clearly seen, as the incident wave is polarized along the y-axis.

Field enhancement factors observed in Raman scattering from molecules adsorbed on nanosurfaces are even larger. The intensity of Stokes component in Raman scattering is proportional to the square of dipole momentum on that frequency [61]:

$$I(\omega_S) \propto d^2(\omega_S) = \alpha_R^2 L^2(\omega_S) L^2(\omega_L) E_0^2, \tag{21}$$

where α_R is the Raman polarizability of the molecule. Here two enhancement effects are present: first, the incident laser field is enhanced by

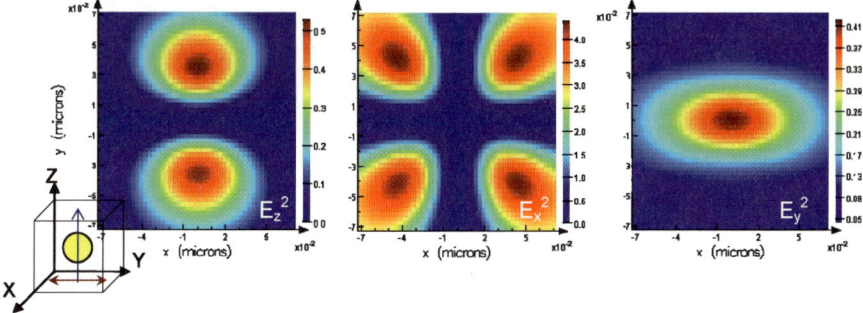

Fig. 12 Optical near-field intensity distribution at the surface of spherical gold nanoparticle with radius of 50 nm, calculated by FDTD technique. The incident field $E(1, 0, 0)$ was polarized along the y-axis; the field monitor plane is perpendicular to the wave propagation direction (z-axis) and located at a distance of 70 nm from the center of the sphere

the nanoparticle according to Eq. 19, and second, an additional emission field of the oscillating molecular dipole is also enhanced by the same nanoparticle. Since usually $\omega_L \simeq \omega_S$, the total enhancement factor becomes $L^4(\omega_L)$ and Eq. 21 becomes $I(\omega_S) \propto \alpha_R^2 L^4(\omega_L) E_0^2$. This demonstrates that nanoparticles act as effective spatial energy redistributors. Raman scattering methods can reach single molecular detection limits by utilizing the surface enhancement discussed above. Methods of nanostructuring by photopolymerization based on this approach are expected to be developed in the future.

4.5
Cavity Enhancement

Enhancement of the light–matter interaction in a microscopic optical cavity is achieved because light trapped in the cavity has longer effective interaction time with absorbers. For short laser pulses, cavity length exceeding $c\tau_p$ allows avoidance of the interference between the pulses incident and reflected from the mirrors. Spectral selectivity of planar Fabry–Perot cavities can be used to achieve the localization at the resonant wavelength of the cavity.

Purcell has demonstrated theoretically that the photon density of states in a cavity having volume V and quality factor Q increases on resonance in

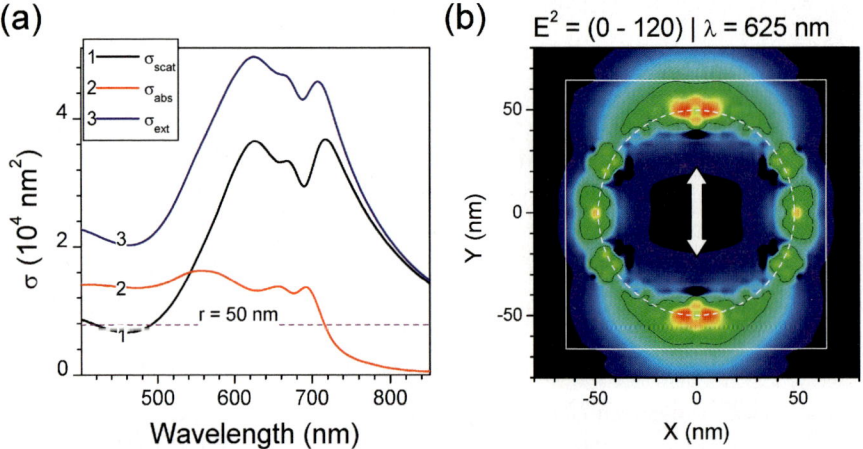

Fig. 13 a Spectra of optical cross-sections of a 50-nm radius gold sphere in a dielectric medium with refractive index of $n = 1.7$: curve 1 σ_{abs} absorption, curve 2 σ_{scat} scattering, and curve 3 $\sigma_{ext} = \sigma_{abs} + \sigma_{scat}$ extinction cross-sections. The *dashed line* marks the geometrical cross-section. **b** Intensity maps of a scattered plane wave $E(0, 1, 0)$ incident on the sphere (the outline is marked in the figure) at 625 nm wavelength (see **a**). The light is incident along the z-axis. The TFSF light source used in modeling the spectra is marked by the *square*. The intensity scale is logarithmic. The *arrow* marks polarization of the incident light. The calculations were done by the FDTD technique [62]

comparison to the free-space value by [64]:

$$f = \frac{3Q\lambda^3}{4\pi^2 V} .$$ (22)

Spontaneous emission and absorption probability is enhanced by the factor f^2 due to the modified optical density-of-states. The Purcell effect becomes increasingly significant for cavities with smaller volume and higher quality factor.

Spherical glass microparticles are simple microcavities having high quality factors. A spherical cavity stores light energy predominantly in a subsurface layer [65, 66], where spatial and spectral concentration of electromagnetic energy takes place. Inelastic light scattering is proportional to the square of the Purcell factor, which for a silica glass spherical microcavity with quality factor $Q = 5 \times 10^9$ becomes $f^2 \simeq 10^{14}$ [67]. Spherical microcavities available in practice have slightly lower Q factors, of the order of 10^8, leading to Purcell factors of $f = 2 \times 10^5$. Consequently, scattering can be enhanced by a factor of $f^2 \simeq 4 \times 10^{10}$. This is much higher than the enhancement achievable with noble metal nanoparticles, approximately 10^6 (Fig. 13).

4.6
Gigantic Second Harmonic Generation

Optical frequency up-conversion, or second harmonic generation (SHG), in nanostructured surfaces can be also considered as a kind of field enhancement [61]. In general, SHG efficiency is proportional to the square of nonlinear polarization $I_{2\omega} \propto [P^{NL}(2\omega)]^2 \propto (\chi^{(2)} E_\omega^2)^2$; here $\chi^{(2)}$ is the second order susceptibility. For a nanostructured surface, the incident field E_ω is transformed to the local field given by Eq. 19, yielding:

$$I_{2\omega} \propto (\chi^{(2)} E_{loc}^2)^2 \propto (\chi^{(2)})^2 L^2(2\omega) L^4(\omega) I_\omega^2 .$$ (23)

Here, two enhancement effects are present: (i) the incident laser field is enhanced in the nanoparticle according to Eq. 19, and (ii) the up-converted field component is also enhanced. The SHG enhancement factors from nanostructured Ag as high as 10^5 (compared to that from atomically flat Ag surface) were observed even for the forbidden case of s-polarized pump beam [61]). This enhancement can be utilized for facilitating TPA and MPA processes.

4.7
Enhanced Absorbance of Molecules Near Metallic Surfaces

Molecules adsorbed on the surface of metals can change their electronic structure, as schematically shown in Fig. 14. Shift and smearing of the mo-

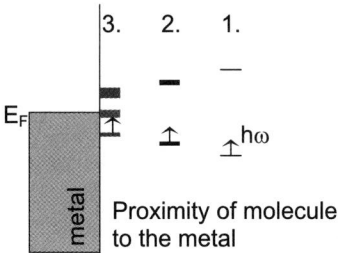

Fig. 14 Schematic depiction of a molecular adsorption on the surface of a metal and corresponding changes in the electronic structure of the molecule, E_F is Fermi level (according to [61])

lecular electronic levels as well as appearance of additional levels may promote both linear and nonlinear absorption.

5
Quantum Lithography

Quantum lithography was introduced recently [68] as a technique providing resolution enhancement beyond the Rayleigh diffraction criterion. This criterion describes fundamental limitation of the resolution achievable with classical optics, for example it defines the minimal achievable lateral size of the focal spot discussed in Sect. 3.1, or spatial period of the two-beam interference field in Sect. 2.2. Quantum lithography exploits correlations between so-called entangled photon-number states. Entangled photons constitute the basis of several other emerging fields (like quantum information processing and quantum cryptography) and have been studied actively during recent years. Their application for quantum lithography was reviewed in [69]. In the next section we will briefly describe basic principles of quantum lithography and outline potential problems on the path toward its practical implementation.

5.1
Principle and Main Requirements of Quantum Lithography

The simplest scheme of quantum lithography is shown in Fig. 15. SPDC is an optical nonlinear process in which one high energy $\hbar\omega_p$ pump photon is split into two (signal and idler) photons having lower energies and nearly perfectly coincident in time. Energy and momentum conservation laws require that:

$$\hbar\omega_p = \hbar\omega_s + \hbar\omega_i \tag{24}$$

$$k_p = k_s + k_i, \tag{25}$$

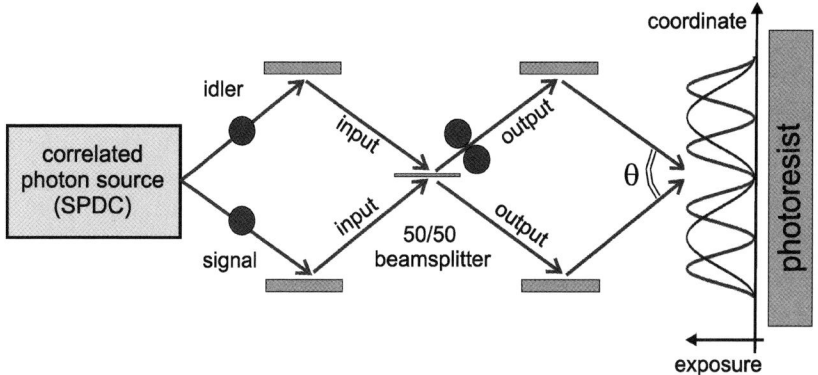

Fig. 15 Schematic implementation of quantum lithography with photon pairs generated in SPDC. Two-photon absorbing photoresist is used as a recording medium. SPDC signal and idler photons are overlapped on a beamsplitter, and exit either of its output ports together. Exposure patterns due to classical and quantum interference are shown in the long and short amplitudes, respectively

where subscripts s and i denote SPDC signal and idler photons, respectively, and magnitudes of the wave-vector $k_{s,i} = 2\pi/\lambda_{s,i}$ are defined by the wavelengths λ corresponding to the photon energies. Energies and propagation directions of the two photons depend on the type of nonlinear interactions and orientation of the nonlinear crystal. In so-called degenerate SPDC, birth-paired photons have nearly the same wavelength $\lambda_i = \lambda_s = 2\lambda_p$. This infers that if a photon pair arrives at a TPA-sensitive absorber atom or molecule at the wavelength of $\lambda_{s,i}$, their perfect correlation may promote a two-photon transition. With uncorrelated photons, a high photon density would be needed in order to achieve close accidental overlap between the photons.

From the birth-paired photons, so-called path-entangled states can be constructed by overlapping them on a non-polarizing 50/50 beamsplitter (the numbers denote transmission and reflection coefficients of the beamsplitter). Two photons having the same polarization and coupled into different input ports of the beamsplitter will interfere and produce a new state that can be described as:

$$\Psi = |2, 0\rangle + |0, 2\rangle . \tag{26}$$

According to this expression, two photons will be always found in one of the output ports of the beamsplitter, while the other port will always be empty. The possibility of one photon in each port is zero, in contrast to what is expected for classical light.

The original proposal of quantum lithography [68] considered exposure of a two-photon absorbing material to two beams of path-entangled photons (e.g., those in the output ports of the beamsplitter), overlapping in space with mutual angle θ. It is helpful to recall that overlap between two classical waves would produce interference fringes with sinusoidal intensity modulation and a spatial period $\Lambda = \lambda/(2\sin(\theta/2))$, where λ is the wavelength. In the limiting case of two counter-propagating waves, $\Lambda = \lambda/2$, the photo-exposure pattern in a recording material photosensitive at this wavelength via linear single-photon absorption will reproduce the intensity pattern of the interference field. Photo-exposure in a material sensitive via TPA at this wavelength will be proportional to the squared intensity, leading to narrowing of the high-exposure fringes, but leaving the fringe period the same as for linear absorption. The main claim of quantum lithography is that two-photon absorption of entangled photon pairs will result in the photo-exposure pattern with doubled spatial frequency, or halved period $\Lambda_e = \Lambda/2$, hence the twice as high spatial resolution. This concept is also applicable to the general case of N entangled photons ($N = 2, 3, ...$) incident on a N-photon absorbing material, resulting in the resolution enhancement by $\Lambda_e = \Lambda/(2N)$.

Despite proof-of-the-principle experimental demonstration of the higher resolution [70], practical recording of periodic patterns using quantum lithography has not been reported yet. The first component needed for successful practical realization of quantum lithography is the intense source of entangled photons. Currently the most accessible and best-studied method of entangled photon generation is the SPDC. Low efficiency of this process results in low entangled photon pair yield, which makes practical studies of quantum lithography difficult. Another required component is the recording medium highly sensitive to photo-excitation via TPA or higher-order absorption. Synthesis of organic dyes intended as TPA-sensitive photo-initiators for photoresists or TPA-excited fluorescence markers for two-photon microscopy is being actively pursued, but their applicability for quantum lithography has not been tested yet.

In the following sections we will give a brief description of essential features of quantum lithography. Fundamental aspects of the entangled photon generation will be mostly omitted from this discussion. The entangled photon source will be treated as a "black box" radiating photon pairs ($N = 2$) at wavelengths and rates close to those obtainable in practice. The main problem addressed here is the optical exposure regime required for quantum lithography, more specifically, under what conditions two-photon exposure and permanent photomodification of photoresists or photosensitive resins is possible.

5.2
Absorption of Entangled Photon Pairs

The crucial prerequisite of quantum lithography is simultaneous absorption of N correlated photons in an N-photon absorbing material. We will discuss TPA of entangled photon pairs along the lines pursued in the literature [71–73]. Entangled photon absorption is often explained by comparison with two-photon absorption of photons randomly arriving from a classical source. Simple probabilistic analysis [72] assumes that two photons couple initial and final states of an atomic media via a virtual state having a lifetime τ. The incident radiation is described by photon flux density, ϕ, or number of photons per unit area per unit time, measured in $cm^{-2} s^{-1}$. As described in Sect. 3.3.2, linear absorption of atoms, molecules, or condensed media via single-photon induced transitions is characterized by the absorption cross-section, σ_{ab}, which has units of area, cm^2. TPA is described by the cross-section $\sigma_{2\gamma}$ and has units of cm^4 s (see below). Efficiency of the absorption process can be characterized by the absorption rate R, or number of transitions per absorber during one second in a unit volume, which has units of s^{-1}. For classical photons the general N-photon absorption rate is determined by the product of the incident flux density and the absorption cross-section for the corresponding process, $R = \sigma_{N\gamma} \phi^N$.

To complete the two-photon transition, the two photons must arrive at the absorber within the virtual state lifetime τ. For classical uncorrelated photons the probability of accidental overlap increases with photon flux density. Therefore excitation by short, tightly focused laser pulses is needed for the TPA. Probabilistic analysis gives the two-photon transition rate:

$$R_2 = \sigma_{2\gamma}\phi^2, \quad \sigma_{2\gamma} = \sigma_{ab}^2\tau, \tag{27}$$

(hence the units of $\sigma_{2\gamma}$ are $cm^4 s^1$. The value of $\sigma_{2\gamma}$ is small for most materials. As mentioned in Sect. 3.3.2, $\sigma_{2\gamma}$ can be conveniently measured in units of $GM = 10^{-50} cm^4$ s. The best TPA absorbing materials have $\sigma_{2\gamma} \sim 10^4$ GM.

Correlation between entangled photons arriving at the absorber in pairs with the total flux density of $\phi/2$ can be described by the entanglement time T_e and entanglement area A_e. For SPDC photons these parameters represent widths of fourth-order temporal and spatial coherence functions [73]. The entanglement time describes the mean time delay between the signal and idler photons. Dispersion in the nonlinear crystal leads to different group velocities of the signal and idler photons having different polarizations, propagation directions, and wavelengths. The entanglement area describes the transverse size of the region in which pairs of photons were generated. The entangled TPA rate is proportional to the joint probability $\xi(T_e)\zeta(A_e)$ that two photons, generated within the time T_e in the area A_e, will arrive at the absorber within the time τ and the transverse area σ_{ab}. Simultaneous presence of two photons at the absorber enhances the probability of TPA. This coincidence is defined by virtue of the generation process, and not by external means like beam focusing.

The probabilistic model gives the entangled pair TPA rate as:

$$R_e = \sigma_e\phi, \tag{28}$$

where the entangled pair absorption cross-section $\sigma_e = \sigma\xi(T_e)\zeta(A_e)/2$. In most cases the virtual state lifetime is much shorter than the entanglement time, and single-photon absorption cross-section is much smaller than the entanglement area. The conditions $\tau \ll T_e$ and $\sigma_{ab} \ll A_e$ allow for the approximations $\xi(T_e) = \tau/T_e$ and $\zeta(A_e) = \sigma_{ab}/A_e$, and consequently:

$$\sigma_e = \frac{\sigma_2}{2A_e T_e} . \tag{29}$$

The entangled pair TPA cross-section σ_e has the same units as the single photon cross-section, cm^2.

Total TPA rate due to the quantum correlation between the entangled photons, and due to their accidental overlap (for example, between photons belonging to different pairs) is a sum of Eq. 28 and Eq. 27:

$$R = R_e + R_2 = \sigma_e \phi + \sigma_{2\gamma} \phi^2 . \tag{30}$$

The two contributions have different dependencies on the photon flux density. For entangled photons the TPA rate is proportional to the flux density, whereas for classical photons it is proportional to the square of the flux density. These dependencies result in two different exposure regimes depending on the photon flux density. The critical photon flux density, ϕ_c, separating these regimes is defined using the condition $R_e = R_2$:

$$\phi_c = \frac{\sigma_e}{\sigma_2} = \frac{1}{2A_e T_e} . \tag{31}$$

At low flux densities, $\phi < \phi_c$, the dominant TPA occurs due to the photon entanglement. Quantum lithography is supposed to work in this regime. At higher flux densities, $\phi > \phi_c$, the dominant TPA is due to the accidental coincidence between photons. In this regime benefits of quantum lithography are lost.

The simple probabilistic model was also confirmed by quantum-mechanical analysis [72]. Quantum-mechanical analysis gave expressions for σ_e and $\sigma_{2\gamma}$ that are very similar to each other. This similarity indicates that TPA of entangled and accidentally coincident photons is essentially identical. Differences between both expressions arise mainly due to the fact the expression for entangled photon TPA cross-section accounts for correlations between the photons: it has a constant factor $1/(T_e A_e)$. Moreover, it includes complex entanglement time-dependent summation coefficients, which may lead to periodic modulation in the $\sigma_e(T_e)$ dependence, or entanglement-induced two-photon transparency.

5.3
Parameters for the Estimates

The expressions given in the preceding section can be used for the estimation of the TPA rates, and help gain the insight into the approximate conditions needed for practical quantum lithography. A direct demonstration of recording by quantum lithography requires use of two entangled photon beams for recording of the periodic pattern with doubled spatial frequency in photoresist, as outlined in Fig. 15. Such demonstration first of all relies on the optical exposure, which is above the threshold exposure dose of the photoresist or other photosensitive material. For practical purposes this dose should be reached within a reasonably short time. We will try to derive approximate numerical estimates corresponding to these requirements, and investigate how they can be met using entangled photons.

In these estimates we consider a model photoresist irradiated by a beam from a model entangled photon pair source. These model systems are assumed to have characteristics close to those actually available. Since exact values of some of their parameters are unknown, or vary in a broad range, preference is given to the values that create more favorable conditions for the implementation of quantum lithography. Examples of such deliberate deviations can be a higher pair yield of the entangled photon source, or a higher TPA sensitivity of the photoresist. The analysis is mainly aimed at evaluating the TPA rates versus the photon flux density for entangled and classical photons.

The main parameters and their values used in the analysis are listed in Table 1. Some of the values are inferred from the reported studies [71–73], and from our own work [74] and references therein. A few comments on some parameters in Table 1 are given below.

Table 1 Basic parameters for the estimates

Parameter	Value	Description
λ	0.8 μm	SPDC photon wavelength
P	10^{-14} – 1 W	Average power in the SPDC beam; the upper limit is unrealistically high
τ_p	100 fs	Pulse length for pulsed SPDC
d	0.8, 50, 0.005 μm	Diameters of the irradiated area
σ_{ab}	10^{-17} cm^2	Typical value of the single photon absorption cross-section for most organic and inorganic materials
R_{pm}	0.1 s^{-1}	Absorption rate at which threshold exposure dose of 1.8×10^{16} photons cm^{-2} is achieved after 1 s exposure time
$\sigma_{2\gamma}$	2.0×10^{-48} cm^4 s = 200 GM	TPA cross-section typical for organic photoinitiators with high TPA sensitivity
T_e	30 fs	Entanglement time
A_e	1.1×10^{-8} cm^2	Entanglement area
ϕ_c	3.0×10^{21} cm^{-2} s^{-1}	Critical flux density
σ_e	2.95×10^{-27} cm^2	Entangled pair absorption cross-section obtained from Eq. 30
f_{pairs}	10^7 s^{-1}	Entangled photon pair rate from the source

Entangled photons emitted by our model source are assumed to have a wavelength of 0.8 μm, at which many cw and pulsed SPDC sources described in the literature operate. The irradiated area also has a diameter of 0.8 μm. Areas comparable with the wavelength can be only achieved in the waist region of a tightly focused beam. In practice, much larger beam diameters are needed for recording of the two-beam interference patterns. The area is thus underestimated, leading to a generally overestimated photon flux density at the given average source power. Transverse intensity variation in the irradiating beam is ignored, therefore its intensity profile has a "top-hat" shape. Likewise, the temporal envelope of the pulsed source is approximated by a square. These two approximations do not influence results of the analysis significantly.

Photon flux density and the average power of the incident beam were varied in a wide range that exceeded the current possibilities. Highest estimates of the photon-pair yields from SPDC sources, given in the literature, are in the megahertz range [74]. Therefore we assume an overestimated pair rate of 10^7 s^{-1}, which corresponds to the photon flux density of 4×10^{15} cm^{-2} s^{-1} for a circular area with diameter of 0.8 μm.

The entanglement time and area depend on the thickness of nonlinear crystal, the type of nonlinear interaction, and pumping conditions. Their chosen values are close to those used in [73]. Together, they yield the critical flux density of $\phi_c = 3 \times 10^{21}$ cm^{-2}. This results in the entangled photon absorption cross-section $\sigma_e = 2.95 \times 10^{-27}$ cm^2. The latter estimate falls between the values obtained earlier from quantum-mechanical calculations for Na (6.0×10^{-30} cm^2) and K_2CsSb (2.6×10^{-25} cm^2) [73].

The model photoresist for quantum lithography is envisaged to have properties approaching those of photoresists currently used in ultraviolet (UV) lithography. In terms of the photomodification mechanism, it is assumed to be a polymeric compound like the popular SU-8 resist. Its single photon absorption cross-section is assumed to be around 10^{-17} cm^2, which is typical for many organic and inorganic molecules, including SU-8 (at 0.365 μm wavelength). Its TPA capability at the wavelength of 0.8 μm is assumed to be close to those of photo-initiator molecules, custom-designed for large TPA cross-sections. One possible example is MBAPB whose TPA cross-section is about 2.0×10^{-48} cm^4 s = 200 GM at 0.8 μm wavelength (see Fig. 9).

Optical exposure must induce permanent photomodification in the photoresist by polymer cross-linking or other mechanisms. Threshold exposure dose is a phenomenological parameter describing the amount of incident radiation required in order to produce permanent photomodification. Both optical absorption and post-exposure effects (like chemical amplification of polymer cross-linking reaction) are included in this parameter. For many photoresists, typical threshold exposure dose attained via single-photon absorption at ultraviolet wavelengths is of the order of 100 mJ cm^{-2}. According to the manufacturer (Microchem), SU-8 requires 100–500 mJ/cm^{-2} at the

356 nm wavelength, depending on the film thickness. A more recent estimate gives threshold exposure dose of 50 mJ cm^{-2} [75]. Here we deliberately lower the threshold exposure dose to 1.0 mJ cm^{-2}, which is equivalent to 1.8×10^{16} photons cm^{-2} (at 356 nm wavelength). Practical applications require reasonably short exposure times. The exposure time of 1 s, at which the exposure dose and the photon flux density have equal values, is a convenient choice. The linear absorption rate is a product of the photon flux density and the linear absorption cross-section. Thus, linear absorption rate $R_{pm} = 1.8 \times 10^{16}$ photons cm^{-2} s^{-1} $\times 10^{-17}$ cm^2 ≈ 0.1 s^{-1} is needed for permanent photomodification of the model resist after 1 s exposure. Similar rates must be also attained for other mechanisms of absorption.

5.4
Two-Photon Absorption Rates Under Continuous-Wave Excitation

The dependencies of TPA and single-photon absorption rates on the photon flux density and average incident power for cw excitation are summarized in Fig. 16. The photon flux density, which is independent of the irradiated area, is given in the bottom abscissa axis. Average power in the beam at a given flux density is indicated in the top abscissa axes. The three axes represent three different diameters of the irradiated area. As discussed above, the entangled photon TPA dominates at photon flux densities below the critical value of $\phi_c = 3.0 \times 10^{21}$ cm^{-2} s^{-1}. This intensity range defines the exposure regime potentially suitable for quantum lithography. The regime described by the photon flux densities higher than the critical value is not suitable for quantum lithography due to the dominance of the uncorrelated photon TPA.

The same plot also includes the dependence of the single-photon absorption rate for the cross-section of 10^{-17}. It can be seen that the entangled photon TPA rate is about ten orders of magnitude lower than the rate of the single-photon absorption. The main reason for this huge difference is the different cross-sections of these processes. As a consequence, the entangled photon rate stays well below the "required" value of R_{pm}. By extrapolating to higher photon flux densities, one can notice that the condition of $R_e \rightarrow R_{pm}$ is reached well above the critical flux density, i.e., outside the exposure regime suitable for quantum lithography. Hence, the entangled photon TPA rate cannot be increased by increasing the photon flux density, which is limited from above by ϕ_c. Photon flux density can be varied by changing the intensity of the source or by focusing/defocusing the beam on the photoresist, which allows adjustment of the relative weights of the entangled and classical TPA rates. In any case, achieving the exposure regime of quantum lithography, $R_2 < R_e$, invariably leads to decreased R_e, and longer exposure time is needed in order to reach the critical dose.

Figure 16 illustrates that the model source providing an entangled photon pair rate of 10^7 s^{-1} and photon flux density of $\approx 4 \times 10^{15}$ cm^{-2} s^{-1} in the area

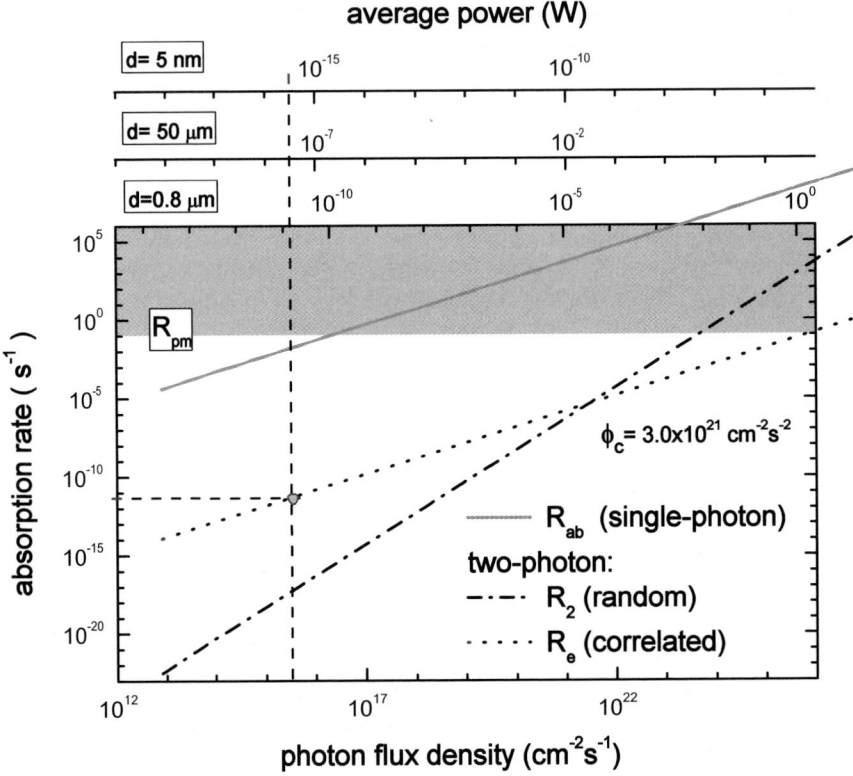

Fig. 16 TPA and single-photon absorption rates versus incident photon flux density calculated using parameters from Table 1. The *bottom abscissa axis* is calibrated in units of
flux density, while the *top abscissa axes* indicate the averages corresponding to the flux
density at three different diameters of the incident beam: 0.8 μm, 50.0 μm and 50 nm. The
gray area outlines the region having absorption rates exceeding the value of $R_{pm} = 0.1$ s^{-1},
which is the estimated average rate needed to photomodify the model photoresist after 1 s
exposure. *Vertical* and *horizontal dashed lines* emphasize the photon flux density and the
entangled photon TPA rates attainable using the model SPDC source

with diameter of 0.8 μm operates in the intensity regime of quantum lithography. However, the low entangled TPA rate $R_e \approx 10^{-11}$ s^{-1} requires a ridiculous
exposure time of the order of 10^{10} s ≈ 300 years in order to reach the threshold exposure. Higher source intensity or a smaller area allow for shorter
exposure times, but due to the limiting critical photon flux density, the shortest available exposure time is still about 30 h.

The top abscissa axes in Fig. 16 compare average powers of the photon
pair source for three diameters of the irradiated area. The lowest axis corresponds to the diameter of 0.8 μm. The middle axis corresponds to the diameter
of 50 μm, which is closer to what would be required for practical recording
of periodic interference patterns. In this case, much higher power levels are

needed to achieve the same flux density. The top axis shows average power for an extremely small irradiated area having subwavelength diameter of 5 nm. Concentration of optical energy in such small area may represent localization of the plasmonic near-field achievable in metallic nanoparticle structures. In this case a lower average power is required for achieving the given flux density. The possibility of exploiting the plasmonic near-field enhancement will be discussed in Sect. 5.6. Notice that according to Fig. 16, even for the smallest irradiated area, the model source with photon pair rate of 10^7 s^{-1} and average power of $\approx 2.5 \times 10^{-12}$ W (at the wavelength of 800 nm) cannot provide entangled TPA rates exceeding $\approx 10^{-7}$ s^{-1}.

5.5
Two-Photon Absorption Rates Under Pulsed Excitation

Pulsed SPDC sources are typically pumped by high repetition-rate lasers. Here the pump pulse duration of 100 fs, and pump repetition rate of $f_{\text{laser}} = 80$ MHz are assumed in correspondence to the typical parameters of frequency-doubled femtosecond Ti : sapphire laser systems. Furthermore, signal and idler pulses of the same length as the pump pulse are assumed. For pulsed excitation it is important to define the photo-excitation regime in more detail. The average number of entangled photon pairs from pulsed and cw sources are usually similar. Therefore an entangled pair rate of 10^7 s$^{-1} = 10$ MHz (corresponding average power is 5 pW) can be used. Since this rate is only one eigth of the pump rate, it is obvious that due to the low SPDC efficiency some pump pulses are "empty" and produce no photon pairs. In practice, SPDC output power and SPDC photon number per pump pulse, n_{pair}, are even lower. A high brightness pulsed SPDC source, proposed and constructed earlier [74], was verified to have an average power below the level of 1 pW. Since $n_{\text{pair}} \ll 1$, it can be assumed that each "non-empty" SPDC pulse carries a maximum of one photon pair. In this case, the average power of the SPDC beam is proportional to the pair generation frequency, f_{pair}, provided that $f_{\text{pair}} < f_{\text{laser}}$. Under these assumptions, only one photon pair is present at the sample, and the instantaneous photon flux density is independent of the average power.

At a first glance, pulsed SPDC may appear advantageous due to its vastly larger instantaneous photon flux density. Temporal localization of SPDC photons within the time interval τ_{p} enhances the instantaneous flux density by the factor of τ_{p}^{-1}. However, the total exposure time (per second) is also reduced to $f_{\text{pair}}\tau_{\text{p}}$. For absorption processes having linear rates, such as single-photon absorption or entangled photon TPA, both factors cancel each other. On the other hand, rates of nonlinear processes like classical TPA become enhanced under pulsed excitation, and additional defocusing/attenuation of the beam might be necessary for achieving the suitable exposure regime.

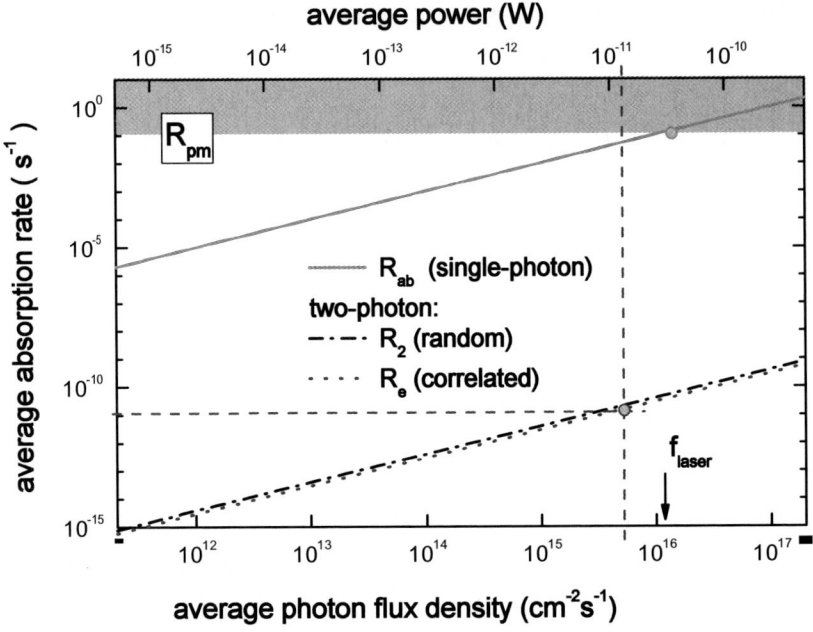

Fig. 17 TPA and single-photon absorption rates versus incident photon flux density calculated using parameters from Table 1 for the case of excitation by pulsed SPDC source. Diameter of the irradiated area is 0.8 μm. The *dashed lines* emphasize the photon flux density and the entangled photon TPA rate provided by the model source

Figure 17 shows same dependencies as Fig. 16 for pulsed excitation and a single size of the irradiated area. As discussed above, the entangled photon absorption rate has remained the same (this can be verified by comparing the corresponding entangled TPA rates). On the other hand, the classical TPA rate has increased by about seven orders of magnitude, and has become nearly equal to the entangled pair absorption rate. Both R_e and $R_{2,\gamma}$ are now linearly dependent on the photon flux density. The latter dependence reflects the fact that instantaneous flux density acting on the absorbers is always constant $(2 \times 10^{21}$ cm^{-2} s$^{-1} > \phi_c)$, while average photon flux density is proportional to the pair arrival frequency. Since a single photon pair has flux density above the critical, this exposure regime is not suitable for quantum lithography. Although the relative weight of the classical TPA rate can be reduced by defocusing the beam, this will also reduce the rate of entangled TPA, just as for cw excitation.

Dependencies of the absorption rates on the irradiation area at a constant pair arrival rate of 10 MHz are shown in Fig. 18. Here R_2 and R_e exhibit different slopes. For areas larger than a certain critical value (which corresponds to the critical flux density, and in this case is very close to the area with diameter of 0.8 μm), the entangled photon TPA dominates the classical TPA. However,

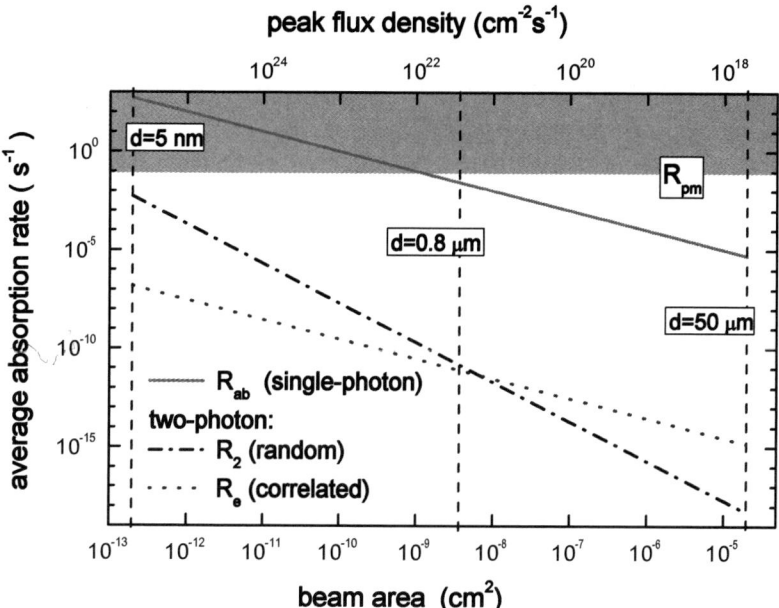

Fig. 18 Dependencies of TPA and single-photon absorption rates on the irradiation area at the constant photon pair arrival rate of 10 MHz. *Vertical dashed lines* emphasize the rates obtained for areas having the indicated diameters

its rate, which was already too small at the diameter of 0.8 μm, becomes reduced even further. On the other hand, decreasing the area invariably leads to the domination of uncorrelated TPA, and unsuitable exposure regime.

5.6
Exploitation of Near-Field Enhancement in Metal Nanostructures

As described in Sect. 4.4, metallic nanoparticles supporting localized surface plasmon (LSP) states can be used for the amplification of optical fields emitted by weak external light sources. Surface plasmons localized on metallic nanoparticles and rough surfaces have their electrical field concentrated near the metal's surface, and field intensity in these small regions becomes orders of magnitude higher than that of the incident wave. Even higher degrees of field localization and intensity can be achieved using two closely spaced nanoparticles separated by distances of a few nanometers only. These "dimers" may consist of low-aspect ratio circular, triangular, or other regularly shaped particles whose smallest separation is typically between their sharp tips. Engineering of size and shape of such particles allows tuning of their collective LSP resonance wavelength; engineering of the interparticle spacing allows changing of the degree of localization and local intensity of the

LSP near-field. It was demonstrated recently by theoretical calculations that at interparticle spacings of about 2 nm, the field is almost completely concentrated in the gap between the particles, and its intensity is about 10^5 that of the incident wave [76].

For our approximate estimates we have assumed a lower degree of localization within an area having a diameter of 5 nm. To verify whether or not such field amplification can be useful for quantum lithography in promoting the entangled photon pair absorption, the case of $d = 5$ nm was included in Figs. 16 and 18. For the cw excitation, Fig. 16 illustrates that a highly localized field allows for a large photon flux density at a lower average incident power. However, the increase in the entangled photon TPA rate remains insufficient. In smaller gaps, stronger field enhancement may even lead to flux densities above the critical value. For the pulsed irradiation, where even larger instantaneous flux densities are obtained due to the short pulses, critical flux density is exceeded even for a single pair of photons, and classical TPA dominates completely. This is demonstrated in Fig. 18. This exposure regime is not suitable for quantum lithography.

5.7
Conclusions

The analysis provided above was based on a physical model known from the literature. Discussion about validity of this model goes beyond the scope of the present study; it is hoped that studies of fundamental quantum optics will be able to provide a more accurate picture. We have also used several additional assumptions and numerical values, which at this moment seem to be a valid approximation of the conditions slightly more favorable for quantum lithography than those actually available. The analysis conducted under these circumstances shows a contradiction between the simultaneous requirements of a sufficiently high irradiation dose and the low incident photon flux density. These requirements can only be met at photon flux densities so low that extremely long exposures would be needed. It is important to stress that this contradiction cannot be resolved by increasing the intensity of the entangled photon source. This conclusion remains valid even if we allow large variations for the parameters used in the analysis.

In these circumstances, the feasibility of quantum lithography depends on the possibility of obtaining absorbers with large cross-sections for the entangled photon TPA, or higher order absorption processes. In fact, the numerical value of σ_e should approach that of σ_{ab}:

$$\sigma_e = \frac{\sigma_2}{2A_e T_e} \longrightarrow \sigma_1 , \tag{32}$$

which requires an increase by about ten orders of magnitude. The entangled photon TPA cross-section depends on the cross-section of classical TPA. Organic molecules optimized for efficient TPA have cross-sections as high as 10^4 GM, i.e., two orders of magnitude larger than assumed in our estimates. However, an increase by a factor of one hundred is clearly insufficient. Further increase in σ_e requires entangled photon sources with improved spatio-temporal coherence, or decreased entanglement area and time in Eq. 32. Reliable estimates of the $A_e T_e$ product are difficult, and at this moment it is not clear how it can be reduced by about eight orders of magnitude. However, if material having a higher value of σ_e can eventually be obtained, the critical flux density would also increase, thus enabling exploitation of high-intensity entangled photon sources in the future.

Besides increasing the process rate, final yield of the entangled photon TPA can be increased by increasing the density of the absorbers. However, the density cannot be increased more than two or three orders of magnitude, since the optical properties of molecular absorbers become different at high densities. Another possible improvement can be achieved by increasing the interaction time between the light and absorbers. In principle, this can be done by using slow light propagation in PBG materials, or localized modes of microcavities.

6
Applications

6.1
Formation of 3D Structures

3D photopolymerization of complex structures, for example PhC templates, usually has a narrow processing window for parameters of resins [56] and resists [57]. Structural integrity of 3D patterns is primarily determined by the elastic properties of photopolymerized materials. In SU-8, the existence of eight cross-linking sites per molecule favors formation of a mechanically strong (with high Young modulus E), but brittle glassy structure. For better cross-linking, preparation procedure of the resist is critical. Spin coating, drying, and annealing should be optimized to obtain the most dense, solvent-free layer for laser recording. This becomes increasingly challenging for SU-8 films thicker than 50 μm, when shrinkage of 3D patterns become observable. Post-exposure baking is the usual step in sample preparation used for enhancement of cross-linking. However, it was recently demonstrated that, even without it, photopolymerization occurs during optical exposure due to considerable local heating which acts as an "optical cure" [57]. The optical cure can be explained by absorption of the thermal emission of thermally non-equilibrated electrons. It seems that spectrally broad continuum-like emission occurs at the IR-wavelengths 2–3 μm where polymers efficiently absorbs.

6.2
Retrieval of 3D Structure

The post-exposure treatment of photopolymerized structures in resin and resist is of critical importance for the final retrieval of 3D structures [77]. Since development of positive or negative resists and resins is a wet process, the damaging effect of capillary forces should be considered during the rinsing and drying of the samples.

For example, capillary pressure of drainage between two planes, separated by distance l_x, would generate a force [78]:

$$F = 2\pi l_z l_y \gamma \cos\theta / l_x ,$$
(33)

which pulls the planes toward each other. Here, γ is the surface tension, θ is the contact angle, and $l_{y,z}$ are the length and height of the planes, respectively. The capillary force determines the minimal spacing of the pattern as a function of the aspect ratio by $d_x \propto \gamma f_{ar}^2$ [79], where the aspect ratio is given by $f_{ar} = l_z / t$ with t being the wall thickness or rod diameter depending on the pattern. Consequently, it is difficult to decrease the spacing of high-aspect ratio structures. Moreover, the bending stiffness of structures depend on their shape and scale, e.g., $\sqrt[3]{E}/\varrho$ and $\sqrt{E}/\varrho$ for planes and rods, respectively [80]. Hence, for the same Young modulus, E, and mass density, ϱ, it is easier to buckle a plane than a rod at a given height.

Columnar structures are more resilient to collapse caused by capillary forces than the vertical planes. In the case of columns [78]:

$$F = 2\pi R l_z \gamma \cos\theta / (1 + D/d) ,$$
(34)

where R is the radius of the pillar and $D = l_y - 2R$ is the gap size. Here, l_y is distance between pillar centers (a period) and d is the projection distance (along the capillary force) from the point of liquid-pillar contact to the pillar's edge ($d \rightarrow 0$ at non-wetting $\theta = \pi/2$, and $d \rightarrow R$ at full wetting $\theta = 0$ conditions).

The critical Young modulus for the pillars to withstand capillary drainage is given by [81]:

$$E_{cr} = \frac{24\gamma l_z^4}{(2R)^3 D^2} .$$
(35)

In photopolymerized structures Young modulus, E, is dependent on the exposure dose and post-exposure treatment, which determine the final degree of cross-linking [82]. Since the thickness of walls and rods in holographic recording is usually controlled by the exposure time, the resultant E value can be different for the thin and thick walls/rods. An additional UV exposure can be applied to improve cross-linking and to increase the Young modulus. Such treatment should even further reduce the feature size of high-fidelity complex patterns.

110 nm

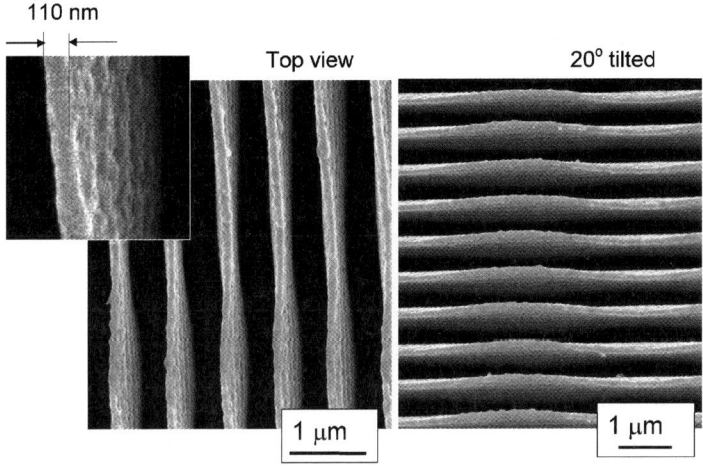

Top view   20° tilted

1 μm   1 μm

Fig. 19 SEM images of 1D pattern recorded holographically by two beam interference in SU-8 and developed using a water rinse to minimize the capillary force of drainage [77]

When standard development of SU-8 in isopropanol was applied to the exposed resist, structures with aspect ratios higher than four could not be obtained without distortion or collapse of planes, despite the low surface tension $\gamma = 21.8$ mJ cm^{-2} of isopropanol. The reason for that was strong wetting ($\theta = 20°$) of SU-8 by isopropanol. Since isopropanol is soluble in water, we added a final rinse in water to the standard development procedure. As a result, high-aspect ratio $f_{ar} = 18$ structures were developed without distortions over the entire exposed region of 700 μm in diameter and entire thickness, as shown in Fig. 19 [77]. Moreover, thickness of the free-standing planes was approximately 110 nm at 0.7 μm period and 3-μm height. This improvement in resolution was achieved due to the hydrophobicity of the SU-8 surface ($\theta = 81°$). Hence, even at the considerably high surface tension of water, $\gamma = 72$ mJ/cm^2, the capillary force (Eq. 33) was approximately half of that in isopropanol. Since water is an environmentally friendly solvent, compatible with most wet processing methods, the proposed technique of capillary force reduction should find wider appeal in wet processing of resists aimed at increased aspect ratio and resolution.

Distortions of patterns due to capillary forces can be avoided altogether by applying supercritical drying (SCD) [83, 84]. In a supercritical liquid, the surface tension becomes negligible ($\gamma \rightarrow 0$) and the capillary force vanishes. However, depending on the solvent, high pressure (1–10 MPa) and elevated temperatures (40–400 °C), may be required [85]. This factor renders this method not universally applicable for wet processing. For example, if SCD was to be used for isopropanol removal from SU-8, temperatures over 235.2 °C and pressures over 4.8 MPa would be needed [85].

6.3
Direct Laser Writing with Ultra-High Resolution

Direct laser writing in SU-8 can achieve resolution of tens of nanometers [86, 87]. SEM image of a linear feature having a diameter of about 30 nm is shown in Fig. 20. This size is comparable with the 2 nm size of the monomer molecules. Fabrication of such structures was accomplished by drawing a pattern of parallel lines spaced by 2 μm from each other by focusing the laser into the the SU-8 layer about 0.7 μm above the substrate surface and translating the sample at a speed of $10 \, \mu m \, s^{-1}$. The laser pulse energy of 3 nJ (measured at the sample) was relatively high, and thick SU-8 rods attached to the substrate were recorded. This first group of lines was intended as a support for the second group of lines, which were drawn immediately after the first group in a perpendicular direction and at the same focusing depth but at a much lower laser pulse energy, 0.3 nJ. At this intensity level, the focal region, where two-photon absorption is effective, did not reach the glass substrate. Consequently, after development, SU-8 nanorods were obtained, suspended in air between the perpendicular thicker rods (note, that since SU-8 remains

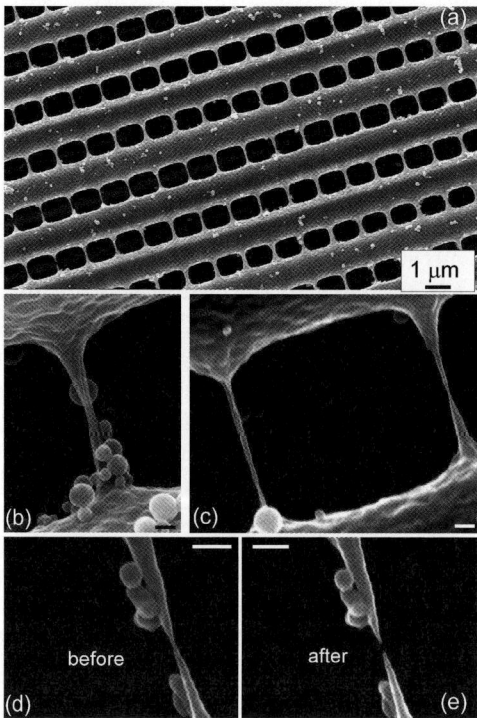

Fig. 20 **a–c** SEM images of nanorods recorded in SU-8, and a single rod before (**d**) and after (**e**) thinning by an electron beam. *Scale bars* are 100 nm (**b–e**)

solid and optically transparent both before and after the exposure, drawing can be performed in arbitrary order). The samples were developed and post-processed following the standard procedure suggested by the SU-8 manufacturer [12]. The resulting structures were sputtered by a 5 nm Pt|Pd and examined by field emission scanning electron microscope (SEM). Figure 20a shows a large-scale image of the sample, which consists of thick SU-8 lines supporting the nanorods.

Figure 20b shows one nanorod in detail; it can be seen that although its cross-sectional diameter varies significantly across the length, the diameter remains smaller than 100 nm, and reaches approximately 30 nm in the waist region. In the image, a number of spherical nanoparticles can also be seen randomly attached to the SU-8 surface. It is likely that they are formed from non-crosslinked monomers during the development. The nanospheres are perfectly shaped, as would be expected for particles formed by surface tension forces. Some nanospheres can be seen attached underneath the nanorod, thus clearly indicating that the nanorods are suspended in air and do not touch the glass substrate.

In general, the shapes and thicknesses of individual nanorods show slight variations, which arise due to SU-8 inhomogeneities and due to the fluctuations of the laser pulse intensity. Some nanorods show tapering at the waist (Fig. 20c), while others have no pronounced waist region and retain an average lateral diameter of approximately 30 nm across their entire length, which corresponds to an aspect ratio of about 20. The average waist diameter of nanorods in the entire pattern was 45 ± 25 nm. It must be stressed that the 30 nm diameter of the nanorod, which constitutes only about 4% of the fabricating laser wavelength ($\lambda/25$ resolution), defines the record-high resolution for two-photon writing of extended features, i.e., rods composed of many overlapping voxels (volume elements). Moreover, these features are suspended in air. For comparison, the highest resolution achieved to date with two-photon exposure (120 nm) was demonstrated by recording isolated voxels in liquid resin directly attached to the glass substrate [88].

It is noteworthy that the nanorods were free of stress. The presence of compressive or tensile stress was tested by cutting through the nanorod using an electron beam. For this purpose, the beam acceleration voltage was set to 10 kV, and rod waist regions narrower than 20 nm were obliterated within 10–30 s time interval. Figures 20d,e show the nanorod before and after the cutting. The width of the gap that developed between the two parts of the nanorod after the cutting was 20 nm, i.e., close to the size of the focal spot of the imaging electron beam. The high acceleration voltage used also contributed to the shortening of the electron de Broglie wavelength and increased the resolution to approximately 5 nm, enabling accurate monitoring of the shape of the nanorod before and after the cutting. After cutting, the shapes of the remaining halves of the rod remained the same, indicating that prior to the cutting the rod was not subjected to significant stress, either due to compression or tension. No buckling of the rod during the thinning (an indication of compressive stress) nor shortening of the remaining halves of the rod (an indication of tensile stress) were observed in our experiments. The lateral shrinkage of up to 10% usually observed in photonic SU-8 microstructures [89] is most probably caused by the capillary forces during the draining stage of development. The sample preparation procedures, thickness of the spin-coated film, and homogeneous solvent evaporation are the most critical for reduction of shrinkage [11].

6.4
Micromechanical Properties

Mechanical treatment of stability developed for macro-objects such as buildings and bridges is also applicable to micro- or nanoobjects such as microtubules, cell membranes, etc [90]. When mechanical objects are small, their functioning can be controlled using weak forces. Usually, weak forces such as van der Walls (dipole–dipole) force or impacts due to Brownian motion become considerable and might be used for sensor applications. Laser microfabrication by photopolymerization of submicrometer-sized objects can find a wider field of applications. Spiral PhC can be fabricated by DLW and holographic lithography techniques (Fig. 21), and used for an optical diode function due to polarization band-gap, i.e., a light of one circular handedness will transmit through a PhC, while the light of opposite handedness will be reflected. Spirals are attractive micromechanical devices and, when small, are responsive to small forces. The total stretch of a spring, f, upon load, P (negative force corresponds to compression), can be calculated as [91]:

$$f = P\left(\frac{R_0^2 L}{GK} - \frac{R_0^2 H_0^2}{GKL}\left(1 - \frac{GK}{EI}\right) + \frac{FL}{AG}\right)$$

$$- P\left(\frac{R_0^2}{3GKL}\left(3 - \frac{2GK}{EI}\right)\left(H^2 + HH_0 + 2H_0^2\right)\right) \tag{36}$$

where E is the Young modulus, R_0 is the initial radius of the coil, H is the variable length of the effective portion of the stretched spring (H_0 is the initial value), L the actual length of the wire of which spring is made, r is the radius of wire, A is cross-sectional area of the wire, K is the torsional stiffness factor ($K = \pi r^4/2$ for a circle; $K = 2.25a^2$ for a square of a side length a), F is the section factor for shear deformation ($F = 10/9$ for a circle and $6/5$ for a square), and I is the inertia momentum of the wire section about a central axis parallel to the spring axis. Hence, f is not a linear function of P. The first term, multiplied by P in Eq. 36, represents the initial rate of stretch, and the second term represents the change in this rate due to modification of the form. This expression is used for high precision calculations of stretch.

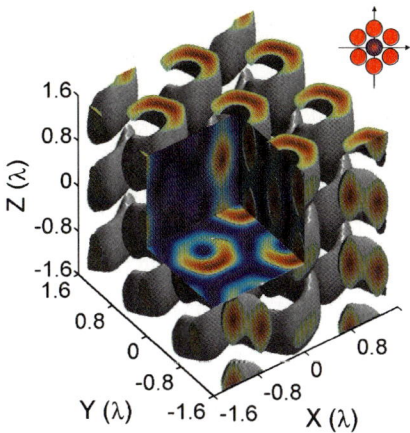

Fig. 21 Spiral structure formed by seven beam interference: central beam is right-handed circularly polarized, and the side beams linearly [32,33]. Simulated by interference of plane waves according to Eq. 2 with side beams comprising an 80° angle with the optical axis (the E-field of the central beam is $E_0 = E_0/\sqrt{2}(1, \pm i, 0)$, where + corresponds to the right-handedness; i $= \sqrt{-1}$)

The spring constant of a helical spring of a circular wire can be calculated by [92]:

$$k = \frac{\pi G r^4}{2R^2 H},\tag{37}$$

where G is the shear/rigidity modulus, $G = E/[2(1 + v)]$ ($v \simeq 0.2$ is the typical value of Poisson's ratio in glasses [92]). A typical set of parameters of spirals polymerized in a SU-8 resist [24, 26, 93] is: coil radius $R = 750$ nm, wire diameter $2r = 500$ nm (this is an approximation since the polymerized wire has an elliptical cross-section), vertical pitch of spiral $c = 2.64$ µm, number of vertical periods $n = 12$ (the total height of structure $H \simeq 31.7$ µm), and $E \simeq 4$ GPa for a fully cross-linked SU-8. Using these, one gets the spring constant $k = 0.57$ N/m from Eq. 37. The load of $P = 1$ mN should generate compaction of a 32-µm tall structure by approximately $f = 12$ µm according to Eq. 36 (here, we consider that the load P is equally distributed over the number of independent springs within the footprint of a 30-µm diameter punching press, and that Poisson's ratio is $v = 0.2$). Hence, a µN force would cause a sub-µm length change detectable by optical means. This might make it possible to tune the stop-band wavelength of a photopolymerized structure. The values of spring constants photopolymerized in resin by femtosecond direct laser writing are not consistent (a difference of five orders of magnitude was reported [94]) with theoretical predictions, most probably due to the unknown Young modulus of the springs. The elastic properties of photopolymerized materials are dependent on cross-linking, which can have a complex dependency on writing conditions.

7
Conclusions and Outlook

We have described principles, methods, and techniques allowing one to exercise a delicate control over the optical fields, and application of these fields for optical fabrication of micro- and nanostructures in photoresists and resins. Microstructuring of materials using lasers promises the availability of relatively easy, simple, and low-cost fabrication processes in the future. As always, there are many challenges remaining to be resolved. The main problem faced by laser lithography is that this optical technique is expected to produce higher-than-optical resolution. One has therefore to use additional mechanisms, like optical nonlinearities, plasmonic localization, or quantum lithography in order to satisfy this demand. Some of these enhancements have already been put to work and have produced spectacular results. Some of them, like quantum lithography, are still being pursued theoretically in expectation of a breakthrough that would allow a more practical work.

Acknowledgements We are indebted to our colleagues, coauthors, and students for numerous discussions on various aspects of laser microfabrication. We acknowledge the discussions and insights of Profs. S. John and G. Ozin on the properties of spiral PhC and for Si-infiltration of SU-8 spiral structures. We also acknowledge the comments of Prof. V. Datsyuk on field enhancement effects.

References

1. Yablonovitch E (1987) Inhibited spontaneous emission in solid-state physics and electronics. Phys Rev Lett 58:2059–2062
2. John S (1987) Strong localization of photons in certain disordered dielectric superlattices. Phys Rev Lett 58(23):2486–2489
3. Strickler JH, Webb WW (1990) Two-photon excitation in laser scanning fluorescence microscopy. SPIE Proc 1398:107–118
4. Denk W, Strickler J, Webb WW (1990) Two-photon laser scanning fluorescence microscopy. Science 248:73–76
5. Strickler JH, Webb WW (1991) Three-dimensional optical data storage in refractive media by two-photon point excitation. Opt Lett 16:1780–1782
6. Wu ES, Strickler JH, Harrell WR, Webb WW (1992) Two-photon lithography for microelectronic application. SPIE Proc 1674:776–782
7. Masters BR (ed) (2003) Selected Papers on multiphoton excitation microscopy. SPIE, Bellingam
8. Bhawalkar J, He GS, Prasad PN (1996) Nonlinear multiphoton process in organic and polymeric materials. Rep Prog Phys 59:1041–1070
9. Göppert-Mayer M (1931) Über Elementarakte mit zwei Quantensprün. Ann Phys Lpz 9:273–295
10. Maruo S, Nakamura O, Kawata S (1997) Three-dimensional microfabrication with two-photon-absorbed photopolymerization. Opt Lett 22:132–134

11. Seet KK (2006) Fabrication of 3D spiral strucure photonic crystals by femtosecond laser and their optical characterization. PhD thesis, Research Institute for Electronic Science, Hokkaido University, Sapporo, Japan

12. MicroChem Corp (2007) SU-8 resist product line. http://www.microchem.com/products/su_eight.htm; last visited 28 Sept 2007

13. Webb RH (1996) Confocal optical microscopy. Rep Prog Phys 59:427–471

14. Ashkin A (1970) Acceleration and trapping of particles by radiation pressure. Phys Rev Lett 24(4):156–159

15. Ashkin A (1997) Optical trapping and manipulation of neutral particles using lasers. Proc Natl Acad Sci USA 94(10):4853–4860

16. Misawa H, Juodkazis S (1999) Photophysics and photochemistry of a laser manipulated microparticle. Prog Polym Sci 24:665–697

17. Straub M, Gu M (2002) Near-infrared photonic crystals with higher-order bandgaps generated by two-photon photopolymerization. Opt Lett 27:1824–1826

18. Juodkazis S, Matsuo S, Misawa H, Mizeikis V, Sun AMB, Tokuda Y, Takahashi M, Yoko T, Nishii J (2002) Application of femtosecond laser pulses for microfabrication of transparent media. Appl Surf Sci 197–198:705–709

19. Serbin J, Ovsianikov A, Chichkov B (2004) Fabrication of woodpile structures by two-photon polymerization and investigation of their optical properties. Opt Express 12:5221–5228

20. Takada K, Sun H-B, Kawata S (2005) Improved spatial resolution and surface roughness in photopolymerization based laser nanowriting. Appl Phys Lett 86:071122–071124

21. Qi F, Li Y, Tan D, Yang H, Gong Q (2007) Polymerized nanotips via two-photon photopolymerization. Opt Express 15:971–977

22. Mizeikis V, Seet KK, Juodkazis S, Misawa H (2004) Three-dimensional woodpile photonic crystal templates for infrared spectral range. Opt Lett 29(17):2061–2063

23. Toader O, John S (2001) Proposed square spiral microfabrication architecture for large three-dimensional photonic band crystals. Science 292:1133–1135

24. Seet KK, Mizeikis V, Juodkazis S, Misawa H (2006) Three-dimensional horizontal circular spirals photonic crystals with stopgaps below 1 μm. Appl Phys Lett 88(22):221101

25. Seet KK, Mizeikis V, Juodkazis S, Misawa H (2005) Spiral three-dimensional photonic crystals for telecomunications spectral range. Appl Phys A 82(4):683–688. doi:10.1007/s00339–005-3459-y

26. Seet KK, Mizeikis V, Juodkazis S, Misawa H (2006) Three-dimentional circular spiral potonic crystal structures recorded by femtosecond pulses. Non-Crystal J Solids 352(23–25):2390–2394

27. Sun H, Mizeikis V, Juodkazis S, Ye J-Y, Matsuo S, Misawa H (2001) Microcavities in polymeric photonic crystals. Appl Phys Lett 79(1):1–3

28. Maznev AA, Crimmins TF, Nelson KA (1998) How to make femtosecond pulses overlap. Opt Lett 23:1378–1380

29. Kondo T, Matsuo S, Juodkazis S, Mizeikis V, Misawa H (2003) Three-dimensional recording by femtosecond pulses in polymer materials. Photopolym J Sci Tech 16(3):427–432

30. Kondo T, Juodkazis S, Mizeikis V, Matsuo S, Misawa H (2006) Fabrication of three-dimensional periodic microstructures in photoresist SU-8 by phase-controlled holographic lithography. New Phys J 8(10):250. doi:10.1088/1367–2630/8/10/250

31. Pang YK, Lee JCW, Lee HF, Tam WY, Chan CT, Sheng P (2005) Chiral microstructures (spirals) fabrication by holographic lithography. Opt Express 13(19):7615–7620

32. Seet KK, Jarutis V, Juodkazis S, Misawa H (2005) Nanofabrication by direct laser writing and holography (invited paper). Proc SPIE Int Soc Opt Eng 6050(60500S):1–9

33. Misawa H, Juodkazis S (2006) Light forms tiny 3D structures. SPIE newsroom. doi:10.1117/2.1200603.0181. Available from http://spie.org/x8778.xml, last visited: 28 Sept 2007

34. Matsuo S, Kondo T, Juodkazis S, Mizeikis V, Misawa H (2002) Fabrication of three-dimensional photonic crystals by femtosecond laser interference. In: Adibi A, Scherer A, Lin S-Y (eds) Photonic bandgap materials and devices. SPIE Proc 4655:327–334

35. Juodkazis S, Kondo T, Mizeikis V, Matsuo S, Misawa H, Vanagas E, Kudryashov I (2002) Microfabrication of three-dimensional structures in polymer and glass by femtosecond pulses. In: Proceedings ROC-Lithuania Bilateral Conf. Opoelectronics & Magnetic Materials, Taipei, May 25–26, 2002, pp 27–29. Available from http://arXiv.org/abs/physics/0205025, last visited: 28 Sept 2007

36. Juodkazis S, Kondo T, Dubikovski S, Mizeikis V, Misawa SMH (2003) Three-dimensional holographic recording in photo-thermo-refractive glass by femtosecond pulses. In: Weber HP, Konov VI, Graf T (eds) Int Conf Advanced Laser Technologies, ALT-2002. SPIE Proc 5147:226–235

37. Kondo T, Juodkazis S, Mizeikis V, Misawa H, Matsuo S (2006) Holographic lithography of periodic two- and three-dimensional microstructures in photoresist SU-8. Opt Express 14(17):7943–7953

38. Sales TRM (1998) Smallest focal spot. Phys Rev Lett 81:3844–3847

39. Friedman E, Miller JL (2003) Photonics rules of thumb: optics, elctro-optics, fiber optics and lasers. SPIE & McGraw-Hill, New York

40. Vogel A, Noack J, Hütman G, Paltauf G (2005) Mechanisms of femtosecond laser nanosurgery of cells and tissues. Appl Phys B 81:1015–1047

41. Sacks Z, Mourou G, Danielius R (2001) Adjusting pulse-front tilt and pulse duration by use of a single-shot autocorrelator. Opt Lett 26:462–464

42. Kawata S, Sun H-B, Tanaka T, Takada K (2001) Finer features for functional microdevices. Nature 412:697–698

43. Gu M (2000) Advanced optical imaging theory. Springer, Berlin

44. Juodkazis S, Okuno H, Kujime N, Matsuo S, Misawa H (2004) Hole drilling in stainless steel and silicon by femtosecond pulses at low pressure. Appl Phys A 79:1555–1559. doi:10.1007/s00339-004-2846-0

45. Trebino R, O'Shea P, Kimmel M, Gu X (2001) Measuring ultrashort laser pulses just got a lot easier. Opt Photon News, pp 23–25. Available from http://www.physics.gatech.edu/gcuo/OPN/GRENOUILLE6-01.pdf, last visited: 28 Sept 2007

46. Akhmanov SA, Vyslouch VA, Chirkin AS (1988) Optics of femtosecond laser pulses. Nauka, Moscow (in Russian)

47. Shoji S, Kawata S, Sukhorukov AA, Kivshar YS (2002) Self-written waveguides in photopolymerizable resins. Opt Lett 27:185–187

48. Band YB (2006) Light and matter: electromagnetism, optics, spectroscopy and lasers. Wiley, UK

49. Cumpston BH, Ananthavel SP, Barlow S, Dyer DL, Ehrlich J, Erskine LL, Heikal AA, Kuebler SM, Lee IS, McCord-Maughon D, Qin J, Rockel H, Wu MR, Wu XL, Marder SR, Perry JW (1999) Two-photon polymerization initiators for three-dimensional optical data storage and microfabrication. Nature 398:51–54

50. Kuebler SM, Braun KL, Zhou W, Cammack JK, Yu T, Ober CK, Marder SR, Perry JW (2003) Design and application of high-sensitivity two-photon initiators for three-dimensional microfabrication. Photochem J Photobiol A: Chem 158(2–3):163–170

51. Zhou W, Kuebler SM, Braun KL, Yu T, Ober JKC, Ober CK, Perry JW, Marder SR (2002) An efficient two-photon-generated photoacid applied to positive-tone 3D microfabrication. Science 296:1106–1109

52. Serbin J, Egbert A, Ostendorf A, Chichkov BN, Domann RHG, Schulz J, Cronauer C, Frohlich L, Popall M (2003) Femtosecond laser-induced two-photon polymerization of inorganic-organic hybrid materials for applications in photonics. Opt Lett 28:301–303

53. Maruo S, Ikuta K (2000) Three-dimensional microfabrication by use of single-photon-absorbed polymerization. Appl Phys Lett 76:2656–2658

54. Juodkazis S, Horyama M, Miwa M, Watanabe M, Mizeikis AMV, Matsuo S, Misawa H (2002) Stereolithography and 3D micro-structuring of transparent materials by femtosecond laser irradiation. In: Panchenko VY, Golubev VS (eds) Seventh international conference on laser and laser-information technologies. SPIE Proc 4644:27–38

55. Witzgall G, Vrijen R, Yablonovitch E, Doan V, Schwartz B (1998) Single-shot two-photon exposure of commercial photoresist for the production of three-dimensional structures. Opt Lett 23:1745–1747

56. Miwa M, Juodkazis S, Kawakami T, Matsuo S, Misawa H (2001) Femtosecond two-photon stereo-lithography. Appl Phys A 73(5):561–566

57. Seet KK, Juodkazis S, Jarutis V, Misawa H (2006) Feature-size reduction of photopoly-merized structures by femtosecond optical curing of SU-8. Appl Phys Lett 89:024106

58. Misawa H, Juodkazis S (eds) (2006) Three-dimensional laser microfabrication: fundamentals and applications. Wiley, UK

59. Iwase Y, Kamada K, Ohta K, Kondo K (2003) Synthesis and photophysical properties of new two-photon absorption chromophores containing a diacetylene moiety as the central π-bridge. Mater J Chem 13:1575–1581

60. Bhardwaj VR, Simova E, Rajeev PP, Hnatovsky C, Taylor RS, Rayner D, Corkum PB (2006) Optically produced arrays of planar nanostructures inside fused silica. Phys Rev Lett 96:057404

61. Akcipetrov OA (2001) Gigantic optically nonlinear effects on the surface of metals. Soros Education J 7(7):109–116 (in Russian)

62. Juodkazis S, Misawa H, Vanagas E, Li M (2006) Thermal effects and breakdown in laser microfabrication. In: Online Proc LAMP2006: 4th int congress on laser advanced materials processing, Kyoto, 16–19 May, 2006. JLPS, Osaka, pp 06–60

63. Kazansky PG, Inouye H, Mitsuyu T, Miura K, Qiu J, Hirao K, Starrost F (1999) Anomalous anisotropic light scattering in Ge-doped silica glass. Phys Rev Lett 82:2199–2201

64. Purcell EM (1946) Spontaneous emission probabilities at radio frequencies. Phys Rev 69:681

65. Datsyuk VV, Juodkazis S, Misawa H (2005) Comparison of the classical and quantum rates of spontaneous light emission in a cavity. Phys Rev A 72:025803

66. Datsyuk VV, Juodkazis S, Misawa H (2005) Properties of a laser based on evanescent-wave amplification. Opt J Soc Am B 22(7):1471–1478

67. Datsyuk V (2007) Ultimate enhancement of the local density of electromagnetic states outside an absorbing sphere. Phys Rev A 75:43820

68. Boto A, Kok P, Abrams D, Braunstein S, Williams C, Dowling J (2000) Quantum interferometric optical lithography: exploiting entanglement to beat the diffraction limit. Phys Rev Lett 85:2733–2736

69. Boyd R, Bentley S (2006) Recent progress in quantum and nonlinear optical lithography. Mod J Opt 53:713–718

70. D'Angelo M, Chekhova MV, Shih Y (2001) Two-photon diffraction and quantum lithography. Phys Rev Lett 87:13602

71. Javanainen J, Gould PL (1988) Linear intensity dependence of a two-photon transition rate. Phys Rev A 41:5088–5091

72. Fei H-B, Jost B, Popescu S, Saleh B, Teich M (1997) Entanglement-induced two-photon transparency. Phys Rev Lett 78:1679–1682

73. Lissandrin F, Saleh B, Sergienko A, Teich M (2004) Quantum theory of entangled-photon photoemission. Phys Rev B 69:165317
74. Jarutis V, Juodkazis S, Mizeikis V, Sasaki K, Misawa H (2004) Ultrabright femtosecond source of biphotons based on a spatial mode inverter. Opt Lett 30:317–319
75. Gaudet M, Camart J-C, Buchaillot L, Arscott S (2005) Variation of absorption coefficient and determination of critical dose of su-8 at 365 nm. Appl Phys Lett 88:24107
76. Hao E, Schatz GC (2004) Electromagnetic fields around silver nanoparticles and dimers. J Chem Phys 120:357–366
77. Kondo T, Juodkazis S, Misawa H (2005) Reduction of capillary force for high-aspect ratio nanofabrication. Appl Phys A 81(8):1583–1586. doi: 10.1007/s00339-005-3337-7
78. Israelachvili JN (1992) Intermolecular and surface forces, 2nd ed. Academic, London
79. Namatsu H, Kurihara K, Nagase M, Iwadate K, Murase K (1995) Dimensional limitations of silicon nanolines resulting from pattern distortion due to surface tension of rinse water. Appl Phys Lett 66(20):2655–2657
80. Ashby MF, Gibson LJ, Wegst U, Olive R (1995) The mechanical properties of natural naterials: material I property charts. Proc Soc R Lond A 450:123–140
81. Tanaka T, Morigami M, Atoda N (1993) Mechanism of resist pattern collapse during development process. Jpn Appl J Phys 32:6059
82. Miwa M, Douoka K, Yoneyama S, Tuchitani S, Kaneko YK (2005) Young's module control of the micro cantilever made by micro-stereolithography. In: El-Fatatry A (ed) MOEMS and miniaturized systems V. SPIE, Bellingham, pp 6–13. doi:10.1117/12.589197
83. McHugh MA, Krukonis VJ (1994) Supercritical fluid extraction, 2nd edn. Butterworth-Heinemann, Boston
84. Eckert CA, Knutson BL, Debendetti PG (1996) Supercritical fluids as solvents for chemical and materials processing. Nature 383:313–318
85. Weissberger A (ed) (1986) Organic solvents. Physical properties and methods of purification, 4th edn. Wiley, New York
86. Juodkazis S, Mizeikis V, Seet KK, Miwa M, Misawa H (2005) Two-photon lithography of nanorods in SU-8 photoresist. Nanotechnol 16:846–849
87. Haske W, Chen V, Hales JM, Dong W, Barlow S, Perry SJW (2007) 65 nm feature sizes using visible wavelength 3-D multiphoton lithography. Opt Express 15:3426–3436
88. Sun H, Kawata S (2004) Two-photon photopolymerization and 3D lithographic microfabrication. Adv Polym Sci 170:169–273
89. Deubel M, von Freymann G, Wegener M, Pereira S, Busch K, Soukoulis CM (2004) Direct laser writing of three-dimensional photonic-crystal templates for telecommunications. Nat Mater 3:444–447
90. Takasone T, Juodkazis S, Kawagishi Y, Yamaguchi A, Sakakibara SM, Nakayama H, Misawa H (2002) Flexural rigidity of a single microtubule. Jpn Appl J Phys 41(5A):3015–3019
91. Young WC, Budynas RG (2002) Roark's formulas for stress and strain, 7th edn. McGraw-Hill, Boston
92. Elmore WC, Heald MA (1985) Physics of waves. Dover, New York
93. Seet KK, Mizeikis V, Matsuo S, Juodkazis S, Misawa H (2005) Three-dimensional spiral – architecture photonic crystals obtained by direct laser writing. Adv Mat 17(5):541–545. doi:10.1002/adma.200401527
94. Sun H, Takada K, Kawata S (2001) Elastic force analysis of functional polymer submicron oscillators. Appl Phys Lett 79:3173–3175

Subject Index

Printing: Krips bv, Meppel, The Netherlands
Binding: Stürtz, Würzburg, Germany